The Humanities in Architectural Design

Offering considerations of the impact which the humanities have had on the processes of architecture and design and subsequently defending their continued relevance, this book explores the issues from both historical and contemporary perspectives.

Written by leading academics in the fields of history, theory and philosophy of design, the essays draw profound meanings from cultural practices and beliefs (religious, mythic, poetic, political and philosophical). The traditional interrelationships between word and image, narrative and space have been altered by the growing dominance of digital media on design and the associated influences of globalization and mass consumption. The chapters here consider how the growing monopoly of technological development in our culture is shaping architectural discourse and how emerging technologies can usefully contribute to a deeper understanding of our design culture.

Quite how we can restore – or indeed radically transform – the traditional dialogue between intellectual enquiry in the humanities and design creativity is at the core of this timely and important book.

Soumyen Bandyopadhyay is Professor of Architecture and Design at Nottingham Trent University and has previously taught at Liverpool University.

Jane Lomholt is a Senior Lecturer in Architecture at the University of Lincoln where she is head of History and Theory.

Nicholas Temple is Professor of Architecture at the University of Lincoln, having previously taught at the University of Liverpool, the University of Pennsylvania and Nottingham University.

Renée Tobe is Architecture Field Leader at the University of East London.

The Humanities in Architectural Design

A contemporary and historical perspective

Edited by
**Soumyen Bandyopadhyay, Jane Lomholt,
Nicholas Temple and Renée Tobe**

Routledge
Taylor & Francis Group
LONDON AND NEW YORK

First published 2010
by Routledge
2 Park Square, Milton Park, Abingdon, Oxon, OX14 4RN

Simultaneously published in the USA and Canada
by Routledge
270 Madison Avenue, New York, NY 10016

Routledge is an imprint of the Taylor & Francis Group, an informa business

© selection and editorial material, Soumyen Bandyopadhyay, Jane Lomholt, Nicholas Temple and Renée Tobe; individual chapters, the contributors

Typeset in Univers by
Saxon Graphics
Printed and bound in Great Britain by
TJ International Ltd, Padstow, Cornwall

All rights reserved. No part of this book may be reprinted or reproduced or utilised in any form or by any electronic, mechanical, or other means, now known or hereafter invented, including photocopying and recording, or in any information storage or retrieval system, without permission in writing from the publishers.

British Library Cataloguing in Publication Data
A catalogue record for this book is available from the British Library

Library of Congress Cataloging-in-Publication Data
The humanities in architectural design : a contemporary and historical perspective / edited by Soumyen Bandyopadhyay ... [et al.].
    p. cm.
  Includes bibliographical references and index.
  1. Architectural design. 2. Humanities. 3. Architecture and society. 4. Design and technology. I. Bandyopadhyay, Soumyen.
  NA2750.H85 2010
  720.1'04—dc22   2009031839

ISBN10: 0-415-55113-7 (hbk)
ISBN10: 0-415-55114-5 (pbk)
ISBN10: 0-203-85944-8 (ebk)

ISBN13: 978-0-415-55113-7 (hbk)
ISBN13: 978-0-415-55114-4 (pbk)
ISBN13: 978-0-203-85944-5 (ebk)

Contents

Illustration credits · vii
Contributors · ix
Acknowledgements · xiii
Introduction · xiv

PART 1
Freedom in the shadow of uncertainty · 3

1   The responsibility of architectural design · 7
      KARSTEN HARRIES

2   *in*Humanities: Ethics "inn" architectural praxis · 16
      LEONIDAS KOUTSOUMPOS

3   Cultivating architects: History in architectural education · 28
      ALEXANDRA STARA

4   The architect as humanist · 36
      LESLIE KAVANAUGH

5   Migration, emancipation and architecture · 43
      JAMES McQUILLAN

PART 2
The spectre of technology · 55

6   Fantasies of the end of technology · 57
      JONATHAN SAWDAY

7   Le Corbusier and the other humanities · 71
      ANDRZEJ PIOTROWSKI

8   Humanist machines: Daniel Libeskind's "Three Lessons in Architecture" · 81
      ERSI IOANNIDOU

9	Draw like a builder, build like a writer ALEX GRAEF AND EDMUND FRITH	91
10	The word made flesh: In the name of the surveyor, the nomad, and the lunatic PETER D. WALDMAN	103

PART 3
Measures of awareness — 119

11	Creative inspirations or intellectual impasses? Reflections on relationships between architecture and the humanities NADER EL-BIZRI	123
12	The human mind and design creativity: Leon Battista Alberti and *lineamenta* NIKOLAOS-ION TERZOGLOU	136
13	Renaissance visual thinking: Architectural representation as a medium to contemplate "real appearance" FEDERICA GOFFI-HAMILTON	147
14	Neoplatonism at the Accademia di San Luca in Rome JOHN HENDRIX	160
15	Who's on first? DONALD KUNZE	172

PART 4
The challenges of instrumental knowledge — 187

16	Architecture as a humanistic discipline DALIBOR VESELY	189
17	The situational space of André Breton's atelier and personal museum in Paris DAGMAR MOTYCKA WESTON	201
18	*L'histoire assassinée*: Manfredo Tafuri and the present TERESA STOPPANI	214
19	Nature choreographed ANNE-LOUISE SOMMER	226
20	The birth of modernity out of the spirit of music? Henry van de Velde and the Nietzsche Archive OLE W. FISCHER	237

Select bibliography	247
Index	255

Illustration credits

The authors and publishers gratefully acknowledge the following for permission to reproduce images in the book. Every effort has been made to contact copyright holders for permission to reprint material. The publishers would be grateful to hear from any copyright holder who is not acknowledged here and will undertake to rectify any errors or omissions in future editions of the book.

PART 1

Figure 1.1 © University of Lincoln
Figure 2.1 © Leonidas Koutsoumpos, collage by the author
Figure 2.2. Photo by Freepal, http://www.flickr.com/freepal/341995265/
Figure 2.3. Collage by the author. Originals of the Russian doll titled 'Lonely peek-a-boo' by Jememoi, http://www.flickr.com/photos/marites/330610550/
Figure 2.4. Photo by Freepal, http://www.flickr.com/photos/freepal/308460780
Figure 5.1 © The Syndics of Cambridge University Library
Figure 5.2 © The J. Paul Getty Museum, Los Angeles. Giovanni Battista Tiepolo, Flight into Egypt (recto); Various Studies (verso), 1725–1735, Pen and brown ink with brown wash over black chalk (recto); faint red chalk (verso), 30.5 × 45.2 cm

PART 2

Figures 7.1, 7.2 and 7.3 courtesy of Fondrem Library, Rice University
Figures 8.1, 8.2, 8.3 and 8.4 © Daniel Libeskind
Figure 9.2 © Jaclyn Holmes
Figure 9.3 © Michael De Wolfe
Figure 10.2 © Alinari/Art Resource New York
Figure 10.3 © Peter Waldman and Chris Genik 1983
Figure 10.4 © Parcel and Peter Waldman 1995
Figure 10.5 © Peter Waldman, Samuel Beall and Justin Walton 2004

PART 3

Figure 11.1 © Heather Miller Photography
Figure 13.1 © Archivio della Fabbrica di San Pietro

Illustration credits

Figure 13.2 © Apollonj Ghetti B. M. et al. 1951
Figure 13.3 © Millon & Lampugnani 1994
Figure 13.4 © Federica Goffi-Hamilton
Figure 13.5 © Courtesy of the BAV
Figures 14.1, 14.2, 14.3 and 14.4 © John Hendrix
Figures 15.1 © Penn State University Library
Figure 15.3 © Donald Kunze

PART 4
Figure 16.1 © Daniel Libeskind
Figure 16.2 © Andrei Leonidov, Family Archive
Figure 16.3 © Board of Trustees of Alfred Beit Foundation and © Dalibor Vesely
Figures 16.4 and 16.5 © Dalibor Vesely
Figure 17.1 © Henri Cartier-Bresson/Magnum Photos
Figure 17.2 © FLC / VG Bild-Kunst, Bonn 2009
Figures 17.3, 17.4 and 17.5b © Dagmar Motycka Weston
Figure 17.5a. Photo by Jacques Faujour © ADAGP, Paris and DACS, London 2006
Figure 18.1 © Nicholas Temple
Figures 19.1, 19.2, 19.3 and 19.4 © Anne-Louise Sommer 2007
Figure 20.2 © Klassik Stiftung Weimar, Goethe-Schiller Archiv
Figure 20.3 © Paul Kühn, Das Nietzsche-Archiv in Weimar 1904
Figures 20.4 and 20.5 © Klassik Stiftung Weimar, 1992

Contributors

Soumyen Bandyopadhyay is Professor of Architecture at Nottingham Trent University. He has researched and published widely on aspects of traditional Omani architecture. He currently heads an AHRC-funded study on the nature of modernity in Indian architecture. He has co-authored a book on the Rock Garden in Chandigarh (Liverpool University Press, 2007) and co-edited a book with Nicholas Temple entitled *Thinking Practice* (Black Dog Publishing, 2007).

Nader El-Bizri is Visiting Professor at the University of Lincoln and Research Associate at the Institute of Ismaili Studies, London. He lectures in History and Philosophy of Science at the University of Cambridge, and is affiliated with the CNRS (Paris). He is general editor of the *Epistles of the Brethren of Purity* (OUP), co-editor of a phenomenology series (Kluwer), editor of the *Encyclopedia of Sciences and Religions* (Springer).

Ed Frith is an architect and academic who has developed a research base with Atelier 11 in Greenwich University and Moving Architecture based in Hackney. Atelier 11 has been studying movements in the Thames Gateway through narrative structures, and Moving Architecture has worked on projects from Hackney to China exploring the practice of architecture. Ed was educated at Bristol and Cambridge, undertook a Fulbright at Princeton, Columbia and with Diller Scofidio Architects, developing projects with his partner, choreographer Caroline Salem. He is a Principal Lecturer at Greenwich University where he is Programme Leader for the Diploma in Architecture.

Ole W. Fischer is a theoretician, historian and practitioner of architecture. From 2002–2009 he taught at the Institute for the History and Theory of Architecture of ETH Zurich, where he also obtained his PhD. Since 2009 he has been a research fellow at the Harvard Graduate School of Design.

Federica Goffi-Hamilton is Assistant Professor at the Architecture Department of Carleton University. She has been Assistant Professor at RISD, where she taught studio architectural representation. She is a PhD candidate in Architectural Representation and Education at Virginia Tech under the guidance of Marco Frascari and Paul Emmons (Virginia Tech). She holds a Dottore in Architettura, University of Genoa, Italy.

Contributors

Alex Graef is an architect and Director of Alex Graef Associated Architects Ltd, Programme Director of the Bachelor of Architecture (Part 2) course at University of Lincoln, UK, and director of the Makers' Research Group. His work aims to create and exploit a symbiotic relationship between architectural practice, teaching and independent research. He practices from Lincoln with links to Vienna, Austria, and Bolzano, Italy.

Karsten Harries is a renowned philosopher, an authority on Martin Heidegger and author of numerous publications on the philosophy of art and architecture. These include *The Meaning of Modern Art* (Evanston: Northwestern, 1968); *The Bavarian Rococo Church: Between Faith and Aestheticism* (New Haven: Yale, 1983); *The Broken Frame: Three Lectures* (Washington, DC: CUA Press, 1990); *The Ethical Function of Architecture* (Cambridge, MA: MIT, 1997) and *Infinity of Perspective* (Cambridge, MA: MIT, 2001).

John Hendrix is a Visiting Professor of Architectural History at the University of Lincoln, UK, and teaches classes at the Rhode Island School of Design in the US. He is the author of several books, including *Architecture and Psychoanalysis*, *Aesthetics and the Philosophy of Spirit*, *Platonic Architectonics*, and *The Relation between Architectural Forms and Philosophical Structures in the Work of Francesco Borromini*.

Ersi Ioannidou is an architect and researcher. She studied at the National Technical University of Athens and the Bartlett School of Architecture, University College London, from where she holds a PhD by architectural design. She has taught in Greece and the UK. Her design work has been exhibited in Athens, London, Seoul and Beijing.

Leslie Kavanaugh is both an architect and a philosopher. She is a Senior Researcher in the philosophy of space and time at TU Delft. She is a registered architect in both America and the Netherlands. She recently published *The Architectonic of Philosophy: Plato, Aristotle, Leibniz* (Amsterdam: University of Amsterdam Press, 2007) and, with Arie Graafland (eds.), *Crossover* (Rotterdam: 010 Publishers, 2006).

Leonidas Koutsoumpos is a registered Architect Engineer, who graduated from the National Technical University of Athens. Being awarded a fellowship by the Greek State Scholarships Foundation, he completed his doctoral research in Architecture at the University of Edinburgh. He currently holds the position of adjunct lecturer at the Department of Architecture, University of Patras, in architecture and urban design.

Donald Kunze has taught at Penn State University since 1984, where he is Professor of Architecture and Integrative Arts. His articles and lectures deal with the poetic dimensions of experience. His book on the philosophy of place of Giambattista Vico studies metaphoric imagination and memory in landscape, architecture, and art. He developed a system of dynamic notation, a graphical approach to problems of the boundary in art, architecture, film, and geographical imagination.

Contributors

Jane Lomholt is a Senior Lecturer in architecture at the University of Lincoln where she is head of History and Theory. She is also Honorary Professor at the University of Applied Sciences, Kaiserslautern, Germany. She studied architecture at the School of Architecture in Aarhus, Denmark, and at the University of Portsmouth. Her PhD from the University of Sheffield is entitled *Theme Park Cities and the American Dream*.

James McQuillan is an architect and theoretician who acquired an MA under Joseph Rykwert at the University of Essex, UK, and spent a research year in Rome. He completed his doctoral thesis on Guarino Guarini in 1991 at the University of Cambridge. He taught at the University of Botswana and is now writing *Modern Architecture as Landscape*, a critical study for the scholar, the professional architect and the student.

Dagmar Motycka Weston is an architect who holds postgraduate degrees from the Architectural Association and the University of Cambridge. She teaches architectural history, theory and design at the University of Edinburgh, and specializes in early twentieth-century art and architecture. Publications include chapters in C. Heck and K. Lippincott (eds), *Symbols of Time in the History of Art* (Brepols, 2002), I. B. Whyte (ed.), *Modernism and the Spirit of the City* (Routledge, 2003) and M. Hvattum and C. Hermansen (eds), *Tracing Modernity* (Routledge, 2004). Her current research project focuses on the artist's studio and personal museum as a matrix of creativity.

Andrzej Piotrowski is an Associate Professor at the School of Architecture, University of Minnesota, USA. Co-editor, with Julia Williams Robinson, of *The Discipline of Architecture* (University of Minnesota Press, 2000) and the author of the forthcoming *The Architecture of Thought*. His scholarship focuses on the interconnections between representation, epistemology and design.

Jonathan Sawday is the Walter J. Ong, SJ, Professor of the Humanities at Saint Louis University in Missouri. He is a cultural historian working on the intersection between science, technology, medicine and imaginative literature. His books include (with Thomas Healy) *Literature and the English Civil War* (1990); *The Body Emblazoned* (1995); (with Neil Rhodes) *The Renaissance Computer* (2000); and *Engines of the Imagination* (2007).

Anne-Louise Sommer is Head of Research and Associate Professor since 2004 at the Danish Design School, Copenhagen. She has held positions at several Danish universities and research centres since the late 1980s. She has published a number of books and articles on Danish architecture and design 1851–1939 and Danish modern furniture design and home decoration 1930–1960. She is the author of *De dødes haver* (2003), *Kaare Klint* (2007), and *Den Danske Arkitektur* (2008).

Alexandra Stara is Director of Graduate History and Theory and Course Director of the MA Thinking Building at Kingston University School of Architecture and Landscape. Current projects include co-editing *Curating Architecture and the City* (Routledge, 2009) and curating *Strange Places*, a major exhibition of contemporary urban and landscape photography at the Stanley Picker Gallery, Kingston, September 2009.

Teresa Stoppani (Architect, IUAV, Venice; PhD Architectural and Urban Design, Florence) is Reader in Architecture at the University of Greenwich, London, UK, where she directs the MSc Architectural Studies and the postgraduate Architecture Histories and Theories programmes. She has a forthcoming book *Paradigm Islands: Manhattan and Venice* (Routledge, 2010).

Nicholas Temple is Professor of Architecture at the University of Lincoln, UK, having previously taught at the University of Liverpool, the University of Pennsylvania and Nottingham University. A graduate of the University of Cambridge, he was a Rome Scholar at the British School at Rome. Publications include *Disclosing Horizons: Architecture, Perspective and Redemptive Space* (Routledge, 2007) and *Thinking Practice: Reflections on Architectural Research and Building Work*, edited with Soumyen Bandyopadhyay (Black Dog Publishing, 2007).

Nikolaos-Ion Terzoglou obtained a Diploma of Architecture (2000), a Master of Science (2001) and a PhD (2005, 2nd ICAR-CORA Award 2007) at the National Technical University of Athens, Greece. He has published a book called *Ideas of Space in the 20th Century* (Athens: Nissos, 2009) and teaches Architectural Theory at the University of Patras.

Renée Tobe studied architecture at the Architectural Association. She has a PhD in Architectural History and Theory from the University of Cambridge. Publications include chapters in *Architecture and Authorship* (Black Dog Press, 2007), *Visual Culture and Tourism* (Berg, 2002), *Disrupted Borders* (Rivers Oram, 1993), and *Ecstatic Antibodies* (Rivers Oram, 1991).

Dalibor Vesely was Director of the MPhil in History and Theory of Architecture at the University of Cambridge and is a visiting professor at the University of Pennsylvania. He is a renowned authority on the philosophy of architecture, in particular the impact of modern technology on traditional world views. His recent book, *Architecture in the Age of Divided Representation* (MIT Press, 2004), enquires into historical and cultural developments in architectural representation and the emergence of a scientific and technological age.

Peter D. Waldman was born in New York City in 1943, studied architecture at Princeton, and served as a Peace Corps Architect in Arequipa, Peru. He then practised and taught at Princeton, Cincinnati, Rice, and since 1992, at the University of Virginia where he is the William R. Kenan Professor of Architecture. A recent Fellow of the American Academy in Rome, his work narrates Spatial Tales of Origin through to strategic Specifications for Construction.

Acknowledgements

The editors wish to thank the following people for their help during the preparation of this book: Kevan Gray, Alistair Wade and Karine Hebel-Deane. Also we would like to extend our sincere thanks to Jennifer Harding and Nick Ascroft, Georgina Johnson and Francesca Ford at Taylor and Francis for their invaluable support.

Introduction

The humanities in crisis?

> It is hypocritical or naïve to defend the humanities as a bulwark against science and technology. The humanities, far from opposing them, are the halo that showers them with meaning and which receives meaning from their efforts.[1]

Ciriaco Moron Arroyo's estimation of the importance, and indeed, relevance, of the humanities in the age of technology provides a fitting point of departure in this investigation of the role of the humanities in architecture. Arroyo's criticism of the all-too-familiar assumption that the humanities serve as a resisting force against the growing domination of science and technology would seem to find some grain of truth in the discipline of architecture, with its interrelated and interdependent fields of enquiry. From the many influences and references that contribute towards the development of design ideas, and their translation into built form, architecture is in every sense a practice that demands negotiation and dialogue between the full spectrum of practical, theoretical, social, political, cultural and technological concerns. It would be worth reminding ourselves of Vitruvius's famous assertion that the architect should be accomplished in a range of skills and have a wide field of knowledge – from moral philosophy to the laws of motion. The Vitruvian perspective is always instructive in this context, more as a rhetorical statement to enhance the social standing of the architect in Augustan Rome than as a reflection of the reality of practice. It provides however little guidance in how we can deal with the multiple and conflicting issues underlying the role of the humanities in architecture today. The theory and practice of architecture have long been subject to differing views about the status of the discipline, as either a "science" or an "art". Against the backdrop of the modern technological age, with the increasing emphasis on material and environmental aspects of sustainability, the scientific outlook has tended to have the upper hand. We need only appreciate the role of new technologies, in the modern age, in addressing social needs and achieving aesthetic effects.

More instructive perhaps is Arroyo's contention that the humanities should be viewed as a "halo" that affirms meaning and purpose to modern science, rather than merely comprising a series of supplementary fields of intellectual enquiry, alongside their scientific counterparts. The assertion is based, of course, on the historical relationship between the humanities and science; that modern science emerged as a "by-product" of its more ancient fields of enquiry, principally philosophy and theology. For many academics and practitioners today, the suggestion of an all-encompassing role for the humanities in our age of information technology would seem to be unfounded, reflecting an outdated perspective of what has become a rapidly changing globalized culture. We need only examine the demise, evident in many aspects of contemporary society, of the various disciplines that constitute the humanities in contemporary culture to appreciate the grounds of this counter-argument. The pervasive nature of popular (media driven) culture and mass consumption, with their obsessions with instant visual gratification and entertainment, has in many instances marginalized, or overshadowed the more reflective activities of the humanities. Architecture is as much a "victim" of this development as most other endeavours, resulting in greater emphasis being placed on the experiential nature of architectural aesthetics over and above questions of meaning.

However, Arroyo also seems to direct us towards the reason of this severance, the curious persistence of an old-fashioned metaphysical distinction between the existence of two completely different kinds of substance: mind and matter. Following different operating principles, as E.G. Slingerland suggests, the mechanistic natural sciences (*Naturwissenschaften*) are subject to *Erklären* or "explanation," while the human sciences (*Geistewissenschaften*) are subject to *Verstehen* or "understanding." This "understanding," the basis of all humanities discourse, takes place through the process which could be seen as an event, "requiring sensitivity, openness and a kind of commitment on the part of one spirit [*Geist*] to another."[2]

Today, by extending this process of "understanding" to encompass the natural sciences and, by implication, technology, the humanities could potentially place themselves in a unique position, willing and able to initiate the process of "consilience" between "understanding" and "explanation" that Arroyo anticipates.[3] To achieve this common ground of continual productive and creative dialogue, while the natural sciences would need to be aware of the limits of objectivism and reductivism they cherish, the humanities should come to terms with how and to what extent the idea of "being human" could be subjected to the rigour of *Erklären*. Reductivism is not the only threat the humanities disciplines face from their scientific counterparts, for constriction introduced from within themselves through certain excesses of post-structuralist and postmodern theorizing have attempted to suffocate the humanities disciplines over the past decades. In extending their open arms towards the sciences and accepting to engage with some of their practices, the humanities would also need to return to their redemptive mandate.

Notwithstanding this apparent contradiction between Arroyo's metaphor of the humanities as a halo, encompassing by their radiance and emanation all scientific and technological advancement, and the decline in the humanities in contemporary culture, it is arguable that the traditions and practices underlying the disciplines of the humanities still provide an important ontological background to our understanding of human endeavours, with their dominant instrumental modes of thinking. The quest for answers to the question of "how," in our relentless search for solutions in technological and scientific advancement, must always be supplemented by posing the question "why." It had been understood, at least until the late eighteenth century, that such twofold questioning ensured a measure of continuity with the past and at the same time served as a mytho-historic reference from which to contemplate new possibilities.

This broadly ethical dimension of the humanities is especially pertinent to architecture, given its historical relationship to the *studia humanitatis* and its deeply embedded associations with social cohesion and religious/political authority, the basis of paradigmatic space. Following this line of thought, the question to what extent we should construe the role of the humanities as a kind of "guardian" to our various roles as architects and architectural thinkers serves as the underlying theme of this study. As Arroyo reminds us, "The greatest value of the humanities lies not in what we can do with them but in what they can do with us."[4]

Orientated thinking

In order to further clarify the relationship between architecture and the humanities it would be helpful, to begin with, to remind ourselves of what constitutes the disciplines of the humanities. Again, Arroyo provides some helpful guidance.

> The humanistic disciplines study man in what makes him different from other entities. They study only man, but as a whole, in philosophy and history. Man is the "talking animal"; language is man as articulated coexistence, consciousness, and self-consciousness, and literature is a unique type of reference to language. And finally man as man is a conscious project. The search for the ultimate meaning of human life and of all reality is theology, as the reflective articulation of religion. The humanistic disciplines are therefore philosophy, history, language, literature and theology.[5]

However, we may question the order and definition of these subjects/disciplines. Arroyo's summary suggests, in a succinct way, that the humanities are not simply a loosely defined collection of distinguishable fields of study, but more crucially an embodiment of the totality of man's sense of order and purpose in the world. This idea of an implicit "wholeness" in the pursuit of knowledge through the humanities derives from the medieval seven liberal arts; these encompass verbal (*trivium*) and mathematical (*quadrivium*) disciplines that defined the curriculum of cathedral schools and early universities.[6] Unlike the traditions of medieval scholasticism, however, knowledge through the *studia humanitatis* was more "pursued" through intellectual enquiry than "revealed" as divine word. This is one of the defining characteristics of Renaissance humanism where the active life (*vita activa*) begins to take precedence over contemplative existence (*vita contemplativa*).[7] Articulated in the writings and ideas of Leon-Battista Alberti and others, this priority formed one of the cornerstones of early architectural theory and gave Renaissance architects, the inheritors of medieval masons and craftsmen, an unprecedented social and cultural status. From Renaissance humanism we witness a gradual but decisive transformation in the understanding of man's capacity to determine his own fate.[8] Crucial to this changing world view, as we know, is the growing awareness of humanity's historicity, of man's historical consciousness, highlighted in antiquarian thought and later archaeology. This led to history itself becoming a branch of the humanities and therefore a legitimate field of intellectual enquiry.

Through the five disciplines expounded by Arroyo, we are given the keys to understanding and reflecting upon our own humanity by recognizing both our limitations and our potential to exceed ourselves. At the same time, the five disciplines indicate various modes of enquiry that give *orientation* and purpose to social, political, religious and cultural order. This sense of directionality of the humanities is important when considering questions of meaning in architecture, where material, spatial, topographical and symbolic aspects combine to convey ritual purpose to human existence. It is this metaphorical exchange between architectural and textual/verbal meanings that has provided the basis of architectural creativity throughout the ages.

Arroyo's distinction between the five disciplines and the human sciences (*Geistewissenschaften*), similar to the humanities but not quite so, Arroyo reminds us, is equally important.[9] These disciplines – anthropology, sociology, ethnography, political sciences and the like – form the immediate ground of negotiation between the humanities and the natural sciences and in locating architecture in its rightful position between the arts and the sciences and between praxis and contemplation. These disciplines aid our ability to simultaneously perceive the specific and the mutable nature of "matter" on the one hand, and the ideal and the universal on the other, which, as Alberto Pérez-Gómez suggests, is at the very basis of the creation of meaning in the world we inhabit. We are capable, as Pérez-Gómez mentions, "of perceiving the ideal in the specific."[10] However, architecture as the recorded palimpsest of all human action is the perennial font to which the humanities also return for replenishment. The seemingly recessive nature of the five disciplines, mediated through the human sciences yet providing a horizon or datum for all considerations pertaining to the specific, are often misunderstood as being absent in architecture.

Concepts are invoked to negotiate the passage between past and present, between architecture and the humanities/technology, and between humanities and the natural sciences. In the art of choosing appropriate and valid concepts lies architecture's ambition to innovate and ensure continuity. The ground of negotiation is being increasingly consolidated through the convergence of the humanities and science disciplines, as is evidenced through recent developments in neuroscience and cognitive science, fundamentally questioning the objectivistic model of the mutually exclusive nature of mind and body. The enactive model of perception, with its parallels in Maurice Merleau-Ponty's phenomenology, emphasizes the image and spatially based nature of concepts and metaphors we employ.[11]

The complexity of the middle ground that architecture intends to negotiate and inhabit, has grown considerably as disparate cultures suddenly find themselves entangled in the ever-expanding web of globalization. While conceptual prerogatives have been understandably cast aside, the challenge is in establishing the means of testing their validity in the face of a worrying expansion of choice. As the one discipline that historically embraces the modernizing divisions between technology and the humanities, architecture provides a unique window into the transformations of culture and society. But such a privileged position leaves open the question of responsibility: how we respond to the demands and expectations of a rapidly changing contemporary culture without losing sight of the traditions of the past. This call for readjustment or renegotiation relies in large measure on our capacity to recognize the interdependence between innovation and tradition, by which the latter is both a "dialogue with the past and an anticipation of the future."[12]

We see this most evidently in recent debates, deliberated by academics and practitioners, about the ethical dimension of contemporary architecture and what we can learn from past traditions as the basis for shaping new practices. From Arroyo's initial statement we can construe that by neglecting the humanities, as precious repositories of intellectual and creative ideas, we are in danger of jeopardizing what is most essential to social cohesion and human well-being, namely, meaning. Historically at least, architecture has provided the ritual and social setting in which questions of meaning, as they pertain to human situations, can be brought to a level of experience.

The shared nature of this experience also alludes to an important role of the humanities in our lives, that of seeking intellectual freedom and liberation from all forms of authoritarianism. In many ways this goal is as relevant today as it was in the early nineteenth century. The essentially interdisciplinary nature of the humanities, as is evidenced from the chapters of this book, seeks intellectual freedom through what Arroyo terms "mediated reflection" on the human condition, and avoiding the pitfalls of dogmatism and populism.[13] However, this freedom is qualified by the sole mandate to understand the human condition, holistically and rigorously, and therefore in a specialized yet interdisciplinary manner.

Summary

The chapters that follow examine, from an array of theoretical and historical perspectives, the role of the humanities in architecture and how the traditional interrelationship between word and image, narrative and space, have been affected by the growing impact of technology and its related fields of media and advertising on the processes of architecture. Based on a selection of papers presented at an international conference, held at the University of Lincoln in November 2007, the chapters draw upon a multitude of references taken from both actual situations (buildings, interiors, landscapes, religious icons, museum/personal collections and dialogues) and various theoretical propositions, influenced by Platonism/Neoplatonism, Arab Aristotelian thought, Renaissance humanism, early modern historiography, phenomenology and hermeneutics. Collectively, these demonstrate the rich and diverse relationships between architectural thought and the various disciplines of the humanities that derive from the *studia humanitatis*. At the same time considerations are made about how the symbolic, analogical and poetic aspects of architectural ideas have, through the humanities, continued to prevail in both the memory and experience of buildings, interiors and landscapes, or through the material legacy of their making in drawings, models, notations, sketches and texts.

These influences are as diverse as the designs they inspire and include religious, political, mythic, poetic, philosophical and historiographical references. But such exchanges between visual and textual material, that have been the basis of architecture and the visual arts for centuries, are being gradually eroded by what might be considered as the increasing marginalization of the humanities in contemporary society in general. Indeed, we need only observe the evident decline in teaching and researching humanities-based subjects in universities today, and conversely the rise of their more "accountable" counterparts in science and technology, to recognize the extent of this wider cultural transformation.

Such a shift in the balance of areas of enquiry, away from the traditional body of subjects derived from the Renaissance *studia humanitatis*, has also brought with it a growing emphasis on specialization at the expense of interdisciplinary work characteristic of the humanities. Architecture stands in an unusual position in this characteristically modern cultural divide, having both "inherited" the traditions and practices of the humanities, in the scope and interrelationship of its various fields of enquiry, and at the same time having become increasingly susceptible to the progressive narrowing and fragmentation of areas of focus or responsibility, evident particularly in professional practice.

Taken from this more general perspective, the subservience of the humanities to technology has led to a blurring, even confusion, between disciplines or cultural practices. We see this in the popular juxtaposition of such terms as culture and industry, as typically applied to design-based activities. In the quest for innovative ideas in design – whether in architecture, landscape architecture or interior design – we are in danger of giving priority to the ephemeral concerns of novelty and visual appeal at the cost of issues of meaning and poetics. Moreover, in the trajectory of technological progress, we are led to assume that creativity can be conveniently equated with production, that how we create is by necessity conditioned by our capacity to visually persuade the user/occupant/consumer. This priority leaves critical questions unanswered about the relationships between information and knowledge, transcription and translation, and how emerging technologies can usefully contribute to a deeper understanding of our design culture. Quite how we can restore, or indeed, radically transform, the traditional dialogue between intellectual enquiry in the humanities and design creativity serves as the overarching theme of this book.

Introduction

At the same time questions are raised about whether the cultural, social and technological changes that have taken place in the modern age have resulted in significant transformations in the traditional role of the various disciplines of the humanities in architecture. In light of the new and increased opportunities for dialogue between the technological and humanistic disciplines, and the immense possibilities offered by the realms of digital technology (once we navigate beyond the obvious seduction of its visual qualities), the role of architecture as the site for such exchange is indeed significant.

The papers selected for this book have been significantly revised and attuned to relate to one of four themes of the text, as summarized by the following titles.

1   Freedom in the shadow of uncertainty
2   The spectre of technology
3   Measures of awareness
4   The challenges of instrumental knowledge

The first part, which is headed by a chapter by Karsten Harries, investigates how the priority of human freedom, characteristic of modernity and inherent in the current era of globalization and mass transit, has significantly shaped our understanding of design as a cultural practice. This priority has left open the question of responsibility in architecture and design and the degree to which our search for limitless possibilities, driven largely by technology, has overshadowed our sense of obligation to larger social and cultural purposes. This tension between freedom and responsibility is initially explored through the relationship between "humanities" and "creativity." Once integrally related and interdependent, the terms have recently acquired divergent meanings that convey contrary views of cultural practice: the humanities as an essentially intellectual and diachronic concern, divorced from preoccupations with originality and innovation, whilst creativity is conceived as a synchronic act antithetical to notions of historical continuity and the "situatedness" of human existence. Against this background the papers raise questions about how the current "lightness" of design can be redeemed from arbitrariness by refocusing on the humanities as a means of reinforcing our dignity as human beings.

The second part investigates pictorial and literary fables, myths, and their relations with technology from the distant past to the present. Jonathan Sawday's introductory chapter examines the peculiar postmodern preoccupation with technological failure. Drawing upon a range of sources, further investigations in this part highlight how technology has enhanced the interrelationship between the humanities and design, through the metaphorical possibilities of machines as narrative structures. However, counter to this, the humanities have become marginalized and obscured through an increasing preoccupation with "know-how" rather than meaning. In a world reliant on technological developments, myths of the machine intertwine with humanity as a flawed prosthesis.

In the third part, which is headed by a chapter by Nader El-Bizri, the authors investigate various intellectual and creative endeavours in the natural sciences, philosophy and architectural theory and how they assist creative development. Although the humanities can grant potential inspiration, and give intellective orientation to design ideas, they may also result in creative impetus, and even encumber us with conceptual/intellectual burdens. This assumption, which is perhaps indicative of the contemporary understanding of the humanities as a source of creative ideas, invariably leads to a tension between text and image, idea and experience. The papers presented here consider how "theories" and their methodologies can inform design creativity and thereby contribute to a deeper sense of meaning in representation.

The final part, headed by a chapter by Dalibor Vesely, explores the evolved role of instrumental knowledge in cultural thought, especially in modernity. Since the eighteenth century, the long tradition of humanist culture has been seriously challenged by the apparent efficiency and adequacy of instrumental representations and models. The chapters presented in this part highlight how architecture and design provide a powerful vehicle for expressing and redefining the role of the humanities in the modern age, against a background of a tendency towards the internationalization of culture and the instrumentalization of life, politics, and power. Each looks in different ways at how architectural representation opens up the possibilities of a sustained and meaningful dialogue between a "past" – variously defined by mytho-historic paradigms or references – and the immediate conditions of contemporary life. Key to this dialogue, as the papers demonstrate, is the role of the humanities in design.

Notes

1. Ciriaco Morón Arroyo, *The Humanities in the Age of Technology*, Washington, DC: The Catholic University of America Press, 2002, back cover.
2. E.G. Slingerland, *What Science Offers the Humanities*, Cambridge: Cambridge University Press, 2008, p. 3.
3. *Ibid.*, p. 2.
4. Arroyo, p. 252.
5. *Ibid.*, p. 18.
6. David Wagner, *The Seven Liberal Arts in the Middle Ages*, Bloomington, IN: Indiana University Press, 1986.
7. On the relation between active and contemplative lives in the Renaissance, see Giuseppe Mazzotta, *Cosmopoiesis: The Renaissance Experiment* Toronto: University of Toronto Press, 2001.
8. Charles G. Nauert Jr., *Humanism and the Culture of Renaissance Europe*, Cambridge: Cambridge University Press, 1995, pp. 8–94.
9. Arroyo, p. 15.
10. Alberto Pérez-Gómez, "Abstraction in Modern Architecture: Some Reflections in Parallel to Gnosticism and Hermeneutics", *VIA, Journal of the Graduate School of Fine Arts, University of Pennsylvania*, Vol. 9, 1988: 71, pp. 71–83.
11. Slingerland, pp. 56–62.
12. Dalibor Vesely, *Architecture and Continuity: Kentish Town Projects 1978–81*, Diploma Unit 1, London: Architectural Association, 1982, p. 8.
13. Arroyo, p. 5.

Part 1

Freedom in the shadow of uncertainty

Contributors to this part explore the role of the humanities in the rapidly changing design environment of extreme uncertainty and arbitrariness. The papers focus on the role of ethics in discharging the responsibility society bestows on architects to create buildings and dwellings. The reflective nature of praxis with ethics embedded at its core, the role of history in expanding and sustaining the ethical responsibility, the reviewed notion of *technē* as the ability to know, and the opportunities posed by migration and the concomitant loss of urban autonomy are discussed by the contributors.

In the first chapter of this part, "The responsibility of architectural design", Karsten Harries discusses the tension that characterises the relationship between the words "creativity" and "humanities". This is especially relevant at a time when it would appear that the role of culture and the humanities has faded into insignificance in our daily lives. Equally, for architectural design creativity, instant simulation is at hand through developments in science and technology, made readily available and immediately applicable. In this context Harries explores the extent to which culture and the humanities could still provide the critical understanding of the limits of freedom we have all grown to accept and cherish. The opportunities of freedom, Harries argues, are being overshadowed by the growing threat of arbitrariness. Citing Milan Kundera, he laments the "unbearable lightness of so much that is being built". Urging designers to be open and responsible, and to step back from the objectifying reason that prevails over the technologically driven nature of our existence, he argues that the importance of the humanities is greater today than ever before.

In "*in*Humanities: Ethics inn architectural praxis", Leonidas Koutsoumpos plays with the word *in*, in an attempt to understand the place of philosophical ethics in contemporary architectural design. Koutsoumpos argues for the need to understand ethics *in* the design activity itself, underscoring architecture's role as a deeply humane endeavour. By pursuing the etymology of the word, Koutsoumpos discovers a connection between *in* and *inn*, an earlier usage richer in meaning that encompassed not only the meaning of "inside" but also the notion "to dwell". Following a

Heideggerian philosophical path, the author unravels two ways of understanding the term *in*: the first is spatial specific, while the second is existential specific and refers to the awareness of being-*in*-the-world. Focusing on the connection between ethics and praxis, the discussion shifts from the dominant theoretical and poetic aspects of architecture to a practical one. Extending the notion of praxis beyond the demands of technological romanticism, he emphasises the ethical term *phronesis* or practical wisdom. Warning against the applied relationship between ethics and architecture, Koutsoumpos argues the deeply embedded nature of ethics in architectural praxis, which, he suggests, should be conceptualised through their *inner* relationship: ethics *inn* architectural praxis.

In "Cultivating architects: The humanities in architectural education", Alexandra Stara proposes to revisit the tradition of the humanities and the underlying theme of a contemporary humanism as the medium for communicating fundamental ideas and demonstrating modes of enquiry, without which, as she argues, both the making and teaching of architecture becomes distorted and problematic. Stara focuses on the role of history in architectural pedagogy, arguing that a historical understanding of architecture is inseparable from its understanding as a creative act. However, she emphasises that the kind of history relevant here has to move away from the established view of teaching history as a systematic collection of information, as well as the equating of truth with fact. In an era of increasing specialisation and difference, there is the pressing need to revisit and redefine the common ground between making and thinking. To fulfil architecture's ethical role and to re-establish its cultural relevance, she emphasises ethical responsibility through a sympathetic and relevant understanding of history.

In "The architect as humanist", Leslie Kavanaugh poses the important question: How does architecture re-situate itself in the humanist tradition? Obviously, humanism has a trajectory that winds its way through history since the time of Erasmus. Humanism, she argues, was not only a plea for rationality, as it is often conceived, but for the self-responsibility of free citizens. Architecture, a profession arising from and existing within this tradition, has always encompassed two streams of thought that can be seen today in the apparent dichotomy of the humanities and the technological sciences; that is to say, the humanist and the master builder. Yet, because the humanist tradition, following Aristotle, exalted man as an animal that reasons, and most specifically an animal that reasons for *himself*, she asks: Should the architect not follow his or her own reason and occupy the place that embraces the free practice of thought and experiment? Emphasising the pervasive notion of *technē* being crucially the "ability" to know rather than "what" to know, she emphasises the humane spatial responsibility of the architect.

In "Migration, emancipation and architecture", James McQuillan reflects on the phenomenon of migration that characterises our cultures at the present time. Migration, as population movement – for work, for tourism and due to strife – McQuillan argues, is of ancient origin, known to the Hebrews and many other races in history. Following Eric Voegelin's treatment of the role of emancipation in Western culture in his *Order and History*, McQuillan asserts migration's emancipatory opportunities. The freedom of the Jews from Egypt, and what Plato called *metaxy*, a realised conflict between divine and mundane orders, were two famous realisations of philosophy arising from ancient experience. Unfortunately, modern architecture failed to focus the attention migration legitimately demanded, beyond a few largely abortive attempts at utopian housing projects. However, contemporary urbanists such as Saskia Sassen have addressed its pivotal importance, highlighted, for example, in tourism as a form of organised migration. The issue of "reverse colonialism" as an emerging cultural, social and political phenomenon is gradually attracting attention from such thinkers and philosophers as Neil L. Whitehead. The resultant loss of autonomy of cities, McQuillan argues, could present a new opportunity for architecture.

Chapter 1

The responsibility of architectural design

Karsten Harries

Introduction

We have come to expect creativity from the architect. What then do the humanities have to contribute to design creativity? There is tension in the question: "Creativity" refers to the ability to invent. It suggests "originality." "Design creativity" would thus seem to call for a certain freedom from the established and accepted ways of doing things, an openness to the challenges presented by an inevitably uncertain future, made exciting, but also troubled, by the new possibilities technology continues to open up. The task of the humanities, on the other hand, is to preserve our shared cultural heritage as expressed in the canonic works of the past. Design creativity demands freedom. The humanities seek to bind freedom to what is most essentially human, thus keeping freedom responsible. But how are we to understand what is "most essentially human"? No longer do we find a ready answer in the works of the ancients, which for so many centuries seemed to provide the humanities with a canon and a foundation. What authority can the humanities still claim?

The tension between "design creativity" and "the humanities" is thus shadowed by the way the latter would seem to possess only a marginal significance in today's world. The humanities may claim to be the custodians of a past that has shaped our values, that has given us our culture, and for centuries provided us with something like a spiritual home. Today that home would seem to lie neglected, if not in ruins, no longer able to provide us with adequate shelter. What importance do the humanities retain in our post-technological, global and market-oriented society? Homer, Virgil, Dante, Shakespeare, Goethe – who still reads them?

George Steiner deplored in a recent lecture the fact that "young Englishmen choose to rank David Beckham high above Shakespeare and Darwin in their list of national treasures" and that "learned institutions, bookstores, concert halls and theaters are struggling for survival in a Europe which is fundamentally prosperous, where wealth has never spoken more loudly." "The fault," he charged, "is very simply ours."[1] But just what is our fault? Who are "we"? Does this "we" include those young Englishmen who care little about Shakespeare and Darwin? Are we to blame them or are they just, like all of us, part of this modern age? Should we blame ourselves for not insisting that our governments spend more on the humanities and culture? What place should the humanities have in a modern democracy? And what place should they have in a university that is truly of today?

The place of design, it would seem, is much less problematic. Just consider such undergraduate programs offered by the Lincoln School of Art and Design: Animation, Conservation and Restoration; Contemporary Decorative Crafts; Contemporary Lens Media; Creative Advertising; Fashion Studies; Fine Art; Furniture; Graphic Design; Product Design; and Illustration and Interactive Design. The engagement of these programs with the world we live in is evident. Not so evident is the relevance of the humanities to these programs. Far more important would seem to be the exciting new possibilities that have been opened up by technological advances that have freed design creativity from long taken-for-granted constraints. Just think of the computer. And do these developments not testify to what is greatest about us humans: our ability to rise above the established and accepted, to meet new challenges, our ability to dream of ever new utopias and to take at least some steps to translate these into reality; in short, the power of self-transcendence that is inseparable from our freedom and has no obvious limit?

In "The Myth of the Man behind Technology" José Ortega y Gasset calls the human being the being that has fallen out of nature, the discontented misfit, the animal that has no home in nature, ever seeking and dreaming of things it had never had. This restless discontent Ortega compares to a love without the beloved, with a "pain that we feel in limbs that we never had." And this discontent he calls what is highest in the human being, just because it is a discontent, because it desires things it has never had. Technology, he suggests, has its origin in such discontent, which demands a new world

> because the real world does not fit us, because it has made us sick. This new world of technology is for us like a gigantic orthopedic apparatus, which you want to create. And all of technology has this wonderful, but – like everything about man – dramatic dynamism and quality of being a marvelous, immense orthopedic creation.[2]

Does design, does our architecture today not possess the same marvelous quality? New possibilities continue to be opened up by science and technology, which have brought with them an ever increasing physical and spiritual mobility, opening up unsuspected ways of dwelling and building. No longer do nature, history, and place determine what or who we have to be. There is a sense in which we have fallen not just out of nature, but out of our history. The crisis of the humanities has its origin in that fall. Is it still possible today to appeal to an essential self that binds freedom?

The responsibility of architectural design

Ortega gave that lecture more than 50 years ago, in 1951, in Darmstadt, addressing the same audience of architects that had just heard Martin Heidegger speak in a much more questioning tone of building and technology in "Building Dwelling Thinking." The technology that to Ortega seemed a marvelous orthopedic invention, Heidegger understood as a threat to our humanity. These two very different judgments, however, do not contradict one another. The freedom that is constitutive of our humanity does indeed harbor within itself a threat to that very humanity, indeed to itself. Presupposing freedom, design creativity, too, has to recognize and confront that danger, struggle to preserve freedom by keeping our freedom from losing touch with our humanity. It is not surprising that this struggle should lead it to look to the humanities for help. But are the humanities still able to offer such help? Has the progress of both freedom and technology that has shaped the spiritual situation of our age not also marginalized the humanities?

There are two challenges Ortega's celebration of technology as a marvelous orthopedic invention must face. The first challenge is presented by the understanding of nature that is a presupposition of our science and technology. Objectifying reason has become the measure of reality. But that understanding of nature has no room for humanity. The second challenge is presented by the progress of freedom that has shaped Western culture and now embraces the globe. This is not to deny the many different kinds of resistance and setbacks with which that progress has met. The difficulty with that progress is that it threatens to outstrip humanity.

Both challenges have the same root: that ability of the human spirit to raise itself, first in thought and then in practice, above whatever place we are assigned by our bodies, to oppose to the perspectives assigned to us by our location in time and space a more objective understanding, to the way things happen to be, other ways in which they might and perhaps should be. Both developments tend towards a marginalization of the humanities. These two challenges have a special relevance for designers, especially for architects, but they have a much broader significance.

To choose, the free subject requires criteria. Imagine having to decide between two courses of action. Either there is a reason to choose one over the other or there is no such reason. In the latter case, if we are not to stand paralyzed before the alternatives, we must "choose" without reason, a "choice" which cannot be distinguished from accident, spontaneity, inspiration – call it what you will. If, on the other hand, there is a good reason, one can ask what makes it so: is that reason good because we chose it, or is its validity independent of our choice? Can such a reason be freely chosen? The original problem reappears. Freedom demands to be bound by what we experience as of decisive significance. To affirm ourselves and our future, we must at least believe that we possess criteria that allow us to make responsible decisions. We may be mistaken in this; our belief may be unwarranted and in bad faith. Still, without such belief there is no genuine choice.

The Enlightenment was confident that human reason could take the place of all the authorities that different revolutions had swept aside, confident that reason and nature were sufficient to lead individuals to the good, the fully humane life. This optimism has supported modernism, as it has supported science, liberal democracy, and international communism. It has also supported modernist architecture from Ledoux to the present. And is it not reason alone that should bind freedom? Where this is accepted and history and religion are subjected to the authority of pure reason, tradition loses its power to place us and the humanities become ever less significant.

But is reason capable of furnishing what is demanded? Many philosophers have seen no great difficulty here. According to Kant freedom perfects itself when it binds itself to the rule of pure practical reason. Kant here presupposes that the rational agent understands himself as a member of a community of rational agents, all of whom demand equal respect. But Kant's rational actor is at some distance from that concrete individual I am, free to ask: Why be moral? This question renders what Kant calls our membership in the Kingdom of Ends problematic, something we can refuse. Kant recognizes the possibility of such a refusal, but considers it radically evil. But the very possibility of evil calls into question any too intimate association of freedom with Kant's pure practical reason. Freedom is open not just to the good, but also to evil. The Enlightenment's faith in the power of reason to build free human beings their spiritual home has been shattered, not only by the horrors of the past two centuries, but by reason itself.

With the eclipse of practical reason, Kantian autonomy had to give way to existentialist authenticity. But, to repeat, divorced from the ability to respond to what already claims us, freedom loses all content and evaporates. Freedom demands such ability to respond. To be sure, our finite freedom is haunted by the possibility of godlike creativity. But this is a temptation. Values or meanings cannot be willed, cannot be freely created; they must be discovered. Without some authority to bind freedom we find ourselves at sea. Contemporary aesthetic production can be cited in support. Let me give you an example.

The better we human beings succeed in asserting ourselves as the masters and possessors of nature, fulfilling that old Cartesian promise, the less will we be able to experience nature as a power that circumscribes and limits who we are and have to be. And the same can be said of our bodies. Are we not on the threshold of becoming designers, not just of buildings, furniture, and clothing, but of our own bodies and selves? Has medical science not given us the means that make remaking ourselves more than just an idle dream, as the French performance artist who has named herself Saint Orlan has attempted to demonstrate with a series of aesthetic events? If successful, these would have shown that, by means of plastic surgery and psychoanalysis, she was indeed able to transform herself so fundamentally that the law would have had to recognize that she had become another person.[3]

> When my operations are finished, I will solicit an advertising agency to come up with a name and logo; next I will retain a lawyer to petition the Republic to accept my new identity and my new face. It is a performance that inscribes itself into the social fabric, that challenges the law, that moves toward a total change of identity.[4]

That what is called here a total change of identity finally makes no sense should be evident: who is that "I" that wants to petition the Republic that it accept the new identity of this "I"?

If the performance invites interpretation as an unsuccessful attempt to commit suicide, it also presents itself to us as an up-to-date version of the old dream of the artist as a second god. Has science not placed us on the threshold of being able to reproduce even human beings? In pursuit of a dream of self-creation what is still understood as art here embraces advertising, the law, and, most importantly, technology. This embrace would reduce supposedly natural and also cultural givens to mere material for technologically aided design. Neither nature, not even our own nature, nor culture can furnish the artist's creativity with a measure to keep such design responsible.

But what is to guide such design? At this point the Cartesian promise of mastery and liberation by means of technology has to give way to dread: unless we ourselves are able to discover the limits or boundaries of the undeniable progress that issued from this promise, must we not lose our way in what knows neither limit nor measure?

In such questioning, underscored by Orlan's "art," our technological civilization's ambiguous discontent with itself finds voice. We may thus lament our malleability, which make us so easily exploited by the advertiser or the demagogue, condemn the rootlessness of our existing, the growing uniformity of a world ever more tightly embraced by technology, mourn the loss of culture, of the kind of *Bildung* associated with the term "humanities." Still, are we to surrender the freedom that finds expression in the art world of today, even if such freedom threatens a loss of humanity?

But is it not the creativity of the artist that is to banish the specter of arbitrariness? Creativity is linked to originality. We may thus want to say that what distinguishes the architect from the mere builder is his original productivity. As Le Corbusier put this point, "Contour and profile are the touchstone of the architect. Here he reveals himself as artist or mere engineer. Contour is free of all constraint."[5] This implies that the artist in the architect is free from all rules.

To link creativity to freedom from all constraints is to characterize it only negatively, as a departure from what has come to be established and accepted, a departure that common sense may well judge nonsense. There are no schools, no text books, no rules that can teach us to be creative in this sense. If creativity is linked to originality, it must be a gift, as Kant insists. Genuine creativity does not strive to be original. The striving for originality substitutes for genuine creativity the pursuit of the interesting. All too much recent architecture would seem to be governed by this pursuit.[6]

The pursuit of the interesting reduces originality to a negation of what has come to be established and accepted. It engages us by presenting us with something unexpected or novel. Its appeal depends therefore on shared expectations that are then disappointed. Thus the normal is boring, the abnormal interesting.

But is the pursuit of the interesting not a distortion of genuine creativity? Does the traditional understanding of the artist as a creator of works that by their inner coherence and integrity defeat arbitrariness not point us in a more promising direction? Invoking the authority of Aristotle, Alexander Tzonis and Liane Lefaivre thus insist that the work of architecture, like every successful work of art, "is a world within the world, 'complete,' 'integral,' 'whole,' a world where there is no contradiction."[7] Given such a conception of the work of architecture as a self-sufficient aesthetic object, all "outside conditions" must be considered "significant obstacles." By its very nature this aesthetic approach is opposed to every contextualism, and that means also to the interesting, which depends on context. Indeed, to the extent that this aesthetic approach governs building, works of architecture will turn a cold shoulder, not only to their neighbors, but to the world that would constrain them with its demands and necessities and to the ground that supports them. Like Ledoux's Shelter for the Agricultural Guards, they are spiritually, if not physically mobile. Such buildings no longer belong to a particular place; they seem ready to travel also in time. There is a sense in which such works seem to have no history.

Milan Kundera spoke of the unbearable lightness of being. Can we not similarly speak of the unbearable lightness of most contemporary aesthetic production? Or of its inhumanity? For to fully affirm ourselves we need to understand ourselves as members, not of Kant's abstract Kingdom of Ends, but concretely, as members of an ongoing historical community, held together by an evolving common sense, long presided over by the humanities. The built environment, too, should answer to that need. Does our architecture answer to that need? Most of the important buildings rising today all over the world, many of them designed by the same small number of star architects, seem mobile in their very essence. Such mobility is indeed demanded by an understanding of the successful work of architecture as first of all an original aesthetic object.

To the extent that a building is experienced as possessing the self-sufficiency demanded of the aesthetic object, it has to be experienced as just happening to occupy this particular place. It thus has to lose the aura that great architecture possesses, precisely because we experience it as firmly placed in its temporal and spatial context. Think of Lincoln Cathedral (see Figure 1.1).

Citing Walter Benjamin, someone is likely to object that to invoke in this way the auratic art and architecture of the past is to nostalgically cling to what cannot and should not be brought back. The challenge demands further consideration. Here I only want to underscore that both, the self-sufficiency demanded of the aesthetic object and the technical reproducibility of the work of art, entail a loss of aura. And the reason is in both cases the same: the loss of transcendence that becomes inevitable when something is experienced as a totally legible and therefore reproducible product of human artifice.

But should what Benjamin called a loss of aura, what Kundera called the unbearable lightness of being, not be understood as an inevitable consequence of humanity's finally coming of age? The fact that Kundera calls this lightness unbearable should, however, make us think. Does freedom indeed become unbearable when it denies us any sense of being firmly placed, physically or spiritually, by something that transcends our ability to produce things?

Figure 1.1
View of Lincoln
Cathedral from
Lincoln Castle

But, to repeat the question: Do such nostalgic invocations of some lost aura not cling to a past that cannot and should not be brought back? Does the emphasis our modern world has placed on the individual not entail an erosion of the power of history to place us? The marginalization of the humanities can be cited as one aspect of this erosion.

Benjamin linked our experience of the aura of a work of art to an appreciation of its uniqueness. Such appreciation he called "inseparable from its being embedded in the fabric of tradition."[8] To tear the artwork out of this context is therefore to destroy its aura.

> Unmistakably, reproduction as offered by picture magazines and newsreels differs from the image seen by the unarmed eye. Uniqueness and permanence are closely linked in the latter, as are transitoriness and reproducibility in the former. To pry an object from its shell, to destroy its aura, is the mark of a perception whose "sense of the universal equality of things" has increased to such a degree that it extracts it even from a unique object by means of reproduction.[9]

Natural objects, too, thus threaten to lose their special aura in the age of mechanical reproduction. The loss of aura can be understood as one of the ways in which what we can call objectification manifests itself. And are not even human beings today in danger of losing that special aura that distinguishes persons from their simulacra? What in principle distinguishes a person from a very complicated robot with a computer brain?

We begin to understand why Benjamin, too, should have found it so difficult to let go of the artwork's special aura, which, he suggests, is not unlike the aura of a person. The experience of another person offers him indeed the paradigm behind all experiences of aura.

> Looking at someone carries the implicit expectation that our look will be returned by the object of our gaze. Where this expectation is met (which, in the case of thought processes can apply equally to the look of the mind and to a glance (pure and simple)), there is an experience of the aura to the fullest extent.[10]

There is to be sure a profound difference between experiencing the gaze of the other, and experiencing the aura of a writer, a composer, a painter, or an architect in one of his or her creations. When I experience the other person, the experience of his or her distinctive aura is the experience of an incarnation of spirit and matter so complete that there is no distance between the two. The mystery of aura is the mystery of such incarnation, which is fully realized when lovers look into each other's eyes: "The person we look at, or who feels he is being looked at, looks at us in return."[11] But something of the sort is present in every experience of aura: to experience the aura of something is to experience spirit incarnated in matter, as if it were another person, capable of speech.

Benjamin no doubt would have us underscore the "as if." "Experience of the aura thus rests on the transportation of a response common in human relationships to the relationship between the inanimate or natural object and man."[12] On this interpretation it is the human subject who invests an essentially mute object with something like spirit of soul. Today a child may still experience rocks and animals as animate, endowed with the power of speech; and fairy tales preserve traces of an older magical experience of the aura of all things. Is a presupposition of our science and technology not a reason that has to render nature mute and meaningless? Such a reason cannot make sense of the phenomenon of aura except as a projection of meaning into matter that as such lacks meaning.

But are human beings not part of nature? What allows us, or Benjamin, in this age of the technical reproducibility, not just of works of art, but increasingly of everything, to hold on to a fundamental distinction between the aura of human beings and the aura of works of art and natural objects? Are not even human beings today in danger of losing that special aura that allows us to experience persons as persons? As Kant recognized, such experience can only be made sense of as an experience of what transcends our understanding. But not just our experience of human beings, our experience of beautiful nature and works of art, too, he thought, opened windows in the edifice raised by science, windows to what alone lets our lives matter.

Let me conclude by returning to our initial question: What do the humanities have to contribute to design creativity? A first answer is to help open windows in the edifice objectifying reason has built, windows to humanity. Such windows are inevitably also windows to transcendence, for freedom transcends whatever objectifying reason can grasp and therefore has no place in its edifice.

But freedom destroys itself without responsibility. This then suggests a second answer: just because design creativity presupposes freedom, it needs the humanities which seek to keep freedom responsible by binding it to what is most essentially human.

Whatever we build or design helps shape the way we are going to live in the future. It may serve and reinforce existing patterns and expectations or it may resist such patterns, perhaps invite us to become more thoughtful. No architect, no designer can escape responsibility for the way we will dwell in the future. But if we are to preserve our humanity we may not allow that objectifying reason that needs to preside over our science and technology to fully determine the shape of the world we live in.

Notes

1   George Steiner, *The Idea of Europe*, introductory essay by Rob Riemen. Tilburg: Nexus Library IV, 2005.
2   José Ortega y Gasset, "The Myth of the Man behind Technology," *Mensch und Raum: Darmstädter Gespräch*, ed. Otto Bartning, Darmstadt: Neue Darmstädter Verlagsanstalt, 1952, pp. 116–17.
3   Linda Weintraub, *Art on the Edge and Over. Searching for Art's Meaning in Contemporary Society*, Litchfield: Art Insights, 1996, pp. 77–83.
4   Saint Orlan, "Intervention," Chapter Nineteen, *The End(s) of Performance*, ed. Peggy Phelan, and Jill Lane, New York: New York University Press, 1997, p. 326.
5   Le Corbusier, "Towards a new architecture: guiding principles," ed. Ulrich Conrads, *Programs and Manifestoes on 20th-Century Architecture*, trans. Michael Bullock, Cambridge, Mass: MIT Press, 1975, p. 61.
6   See Karsten Harries, "Modernity's Bad Conscience," *AA Files*, No. 10, Autumn 1985, pp. 53–60.
7   Alexander Tzonis and Liane Lefaivre, *Classical Architecture. The Poetics of Order*, Cambridge, MA and London: MIT Press, l986, p. 9.
8   Walter Benjamin, "The Work of Art in the Age of Mechanical Reproduction," trans. Harry Zohn, *Illuminations*, New York: Schocken, 1968, p. 223.
9   *Ibid.*
10 Walter Benjamin, "On Some Motifs in Baudelaire," *Illuminations*, p. 188.
11 *Ibid.*
12 *Ibid.*

Chapter 2

*in*Humanities

Ethics in architectural praxis

Leonidas Koutsoumpos

Introduction

> For this is humanism: meditating and caring, that man be human and not inhumane, "inhuman," that is, outside its essence.[1]

This chapter aims to discuss the premise behind the title of this book and the inherent value of its constitutive terms, "humanities" and "design creativity." The fundamental hypothesis of this title is the fact that the humanities have been marginalised in higher education by the monopoly of technological development. This view gives to the humanities an inherent value in the context of the design disciplines and architecture in particular. Moreover, this value is arguably not aesthetic or of any other kind, but primarily ethical. It is exactly this ethical value that I am going to discuss, arguing that it lies *in* the practice of architecture.[2] My argument opposes other views that approach the problems of ethics as an application of a theoretical problem to "real world dilemmas." Often such views end up suggesting norms or rules that can be applied on architecture as external imperatives. But what does the term *ethics* mean in the context of architecture?

Here, I will deliberately avoid the problem of defining ethics[3] by giving two examples that I hope everyone understands as having to do with ethics: the fall of the wall in Berlin, and the building of the wall in Jerusalem. Although I will not analyse them as political or urban/architectural actions, I will illustrate my arguments with constant references to both of these walls. Moreover, I am going to provide the reader with a sharp observation by Maurice Lagueux who stated that architecture constantly raises ethical problems because it "produces the obligatory framework *in*Humanities: Ethics in architectural *praxis*

for social life."⁴ Despite how little or how much we know about the walls in Berlin and in Jerusalem, we can understand that having a wall dissecting or circumscribing a city can cause major distress for the people that live on either side, exactly because of its obligatory character, causing isolation, depriving access, separating families, promoting nationalism and so on. Moreover, I would like to argue that these two examples are paradigmatic metaphors of every architectural activity, which, arguably, can be reduced to two actions: breaking and creating boundaries. Both of those are also paradigmatic because they highlight the two ethical extremes of "human" (the fall of the Berlin Wall as "good") and "inhuman" (the erection of the Jerusalem wall as "bad") practice.

Having provided a preliminary description of ethics, I will proceed by investigating the word "in" in an attempt to understand the place of ethics in contemporary architectural design. By arguing for the need to understand ethics *in* the design activity itself (seen as *praxis*), I see architecture as an *in*herent human activity. But, at the same time, I will challenge the centrality of human rationality as it appears in the tradition of the Enlightenment and I will highlight its *in*human potential.

Defining "in" and its architectural metaphors

Although a small word, "in" can be used in a wide variety of ways that have interesting references to architecture. "In" most commonly appears as a preposition. For example when Descartes says "I am here, seated by the fire, attired *in* a dressing

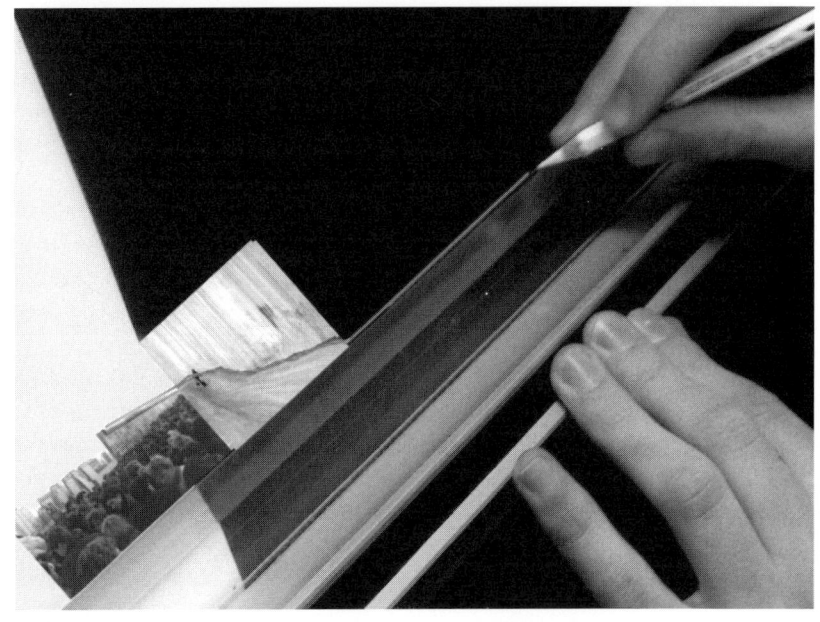

Figure 2.1
The praxis of architecture: Drawing walls as a paradigmatic architectural activity

gown, having this paper *in* my hands"[5] we can understand that he is *in* his study room, his body is *in* his gown (seventeenth-century robe) and he holds a paper *in* his hand. In the context of architecture, the preposition "in" is location specific.[6] According to the Oxford English Dictionary (OED), it is primarily understood in opposition to the preposition "out" (of), where "anything which is *in* a given space is not *out of* it, and *vice versa*."[7] So, in these terms we can understand Descartes' location from its dialectic opposite: he is *not* out of the house (he is not homeless), his body is *not* outside of the robe (he is not naked) and the paper is *not* in the fireplace (as Hume would have liked[8]). This definition, through its opposite, describes the word "in" from the outside. Moreover, the compound "with-in" places emphasis on "the relationship to the limits"[9] in the same sense that West Berlin was with-in East Germany and it was defined by its limits, the wall. In a very similar sense, the simplest definition of humanism finds itself identified by its contrast to its categorical opposite: the animals. Especially the very first notion of Roman humanism constitutes man to be *animal rationale*.[10] In this sense the boundary separating humans and animals is reason, in a very similar sense to Descartes identifying his existence with his ability to think (*cogito ergo sum* = I think therefore I am). Reason and the *cogito* is the wall that keeps us humans *in* humanity.

"In" as within relates to the material extension of the boundaries by being "within" any place or thing.[11] For this it expresses "relation to that which covers, clothes or envelops,"[12] its material, its colour (*in* linen, *in* red). The limits or bounds can be either real and literal (in a box) or virtual and immaterial (in a book, in school). "In" also refers to a condition or state, physical, mental or moral, e.g. *in* doubt, *in* love, or can relate to a manner (way, mode, style, fashion). It can also express reference or relation to something: in reference, in regard to, in the case of, in the matter of, and so on. Furthermore, the Latin preposition "*in*" enters in a number of formal phrases that refer to theology, law, logic or philosophy; for example, *in situ*, meaning in its original place, in position, or *in actu*, meaning "in practice (as opposed to theory or potentiality)."[13] While this paper argues that ethics should be seen as being "in practice," though it is not interested in such a formal way of expressing (*moralia in actu*), it does refer to a notion of practice that, as an envelope, embraces ethics.

The word "in" can also be found as a prefix. The OED, again, argues for an understanding of complicated changes from *in-* to *on-* and sometimes reduced to *a-*. For example, the Gothic word *inliuhtjan*, through the Old High German *inliuhten* and the Old English *onliehtan*, came to mean "to enlighten, to illuminate";[14] our well-known Enlightenment. There is also another prefix that comes from the Latin *in-* which cognates with the Greek α- and the common Teutonic *un-*. This is usually prefixed to adjectives and their derivatives, expressing negation or privation.[15] In this sense "inhuman" is an adjective referring to either a person without the qualities proper or natural to a human being,[16] or a practice that is "brutal, savage, barbarous or cruel."[17] Arguably such cruelties are happening nowadays in Jerusalem by erecting a "monstrous" wall that negotiates one of the most fundamental human needs: "to dwell" or to feel at home. The state of Israel backs up the building of the wall in Jerusalem by using a rhetoric that prioritises the security of the Israeli citizens' homes from being attacked by the Palestinians. Ironically, at the very same time, this wall deprives the Palestinians from their own homes and their feeling of security or "dwelling." Although Martin Heidegger challenges whether the essence of man "lies in the dimension of *animalitas* at all,"[18] humans persistently behave like beasts to each other by employing strategic reason to survive. Architecture and the building of walls participate predominantly in their exercise of "cruelties" in the name of reason.

"In" appears also as a noun and an adjective, but due to the limited space I will choose to focus on its usage as an adverb and a verb. As an adverb, apart from its positional characteristics, it can also express motion or direction, sometimes towards a central point, or most often "from a point without certain limits to a place within these; so as to penetrate or pass into a certain space."[19] "In" as a verb today means "to take in, include, inclose"[20] or to gather into the barn or the stackyard, to harvest or house, and generally to collect. It has etymological references to the adverb mentioned before, which appears to have a common origin in various Common Teutonic languages, like the Old English *in* (and the noun from the Old Frisian, Old Saxon, Old High German *in*) as well as the Middle High German *in*, *în*, and the German *ein*. In Gothic language, when composed with other verbs, "in" becomes *inn-*, a transformation that is also apparent in the Old Norse *inn*. According to the OED, some Old English words like *innian* or *einnian*, "appear to attach themselves in part directly to the adverb inn, partly to be more immediately associated with the derivative, inn."[21] The verb "inn," now rarely in use,[22] used to mean to lodge, house, or the action of finding a lodging. As a noun "inn" used to mean a dwelling place, habitation, abode, lodging; or more simply a house (in relation to its inhabitant).[23] It also refers to a public house for lodging and entertainment of travellers, a kind of hotel, but today "sometimes, erroneously, it means a tavern which does *not* provide lodging."[24]

The reader may find all this etymological archaeology unnecessary and boring, but I think that we are at the threshold of an important discovery. We found

Figure 2.2
No way in (note the ironic label 'Welcome to Jerusalem'!)

that "in" used to have another "n" in it and for this, it used to be richer in meaning. And here, I would also like to argue that over the centuries (especially during those of the Enlightenment), with the reduction from the 'inn' to the 'in' we have lost more than a letter. By using the rhetoric of rationality and efficiency, we are building an inhuman wall in Jerusalem that creates boundaries between our co-humans. As an outcome of our eagerness to become more human, we have lost something that is crucial for the understanding of ethics in architecture (that ethics dwell *in* its practice) and for this we have become more *in*human.

"Inn" and Heidegger: Dwelling on the hyphens of "being-in-the world"

This understanding of "in" – and consecutively "inn" – as dwelling, finds itself at home in the philosophy of Heidegger. Especially in his major work *Being and Time*, Heidegger makes an explicit reference to the notion of "in" as residing.

> "In" is derived from "*innan*" – "to reside," "*habitare*" "to dwell" [*sich auf halten*]. "*An*" signifies "I am accustomed," "I am familiar with," "I look after something."[25]

As the English translators (John Macquarrie and Edward Robinson) inform us in a footnote to the words "to reside" of the above quote, Heidegger himself references back to Jacob Grimm's *Kleinere Schriften*, Vol. VII, pp. 247ff. There Grimm compares various archaic German words that mean "*domus*" and he also mentions that they are similar to the English word "inn," as we saw earlier. He then suggests "a strong verb 'innan', which must have meant either '*habitare*', '*domi esse*', or '*recipere in domum*'."[26] Moreover, Grimm argues that our most familiar notion of "in" as preposition actually derives from the form of the word as a verb, and not the other way round.

Looking back to the original German text of the above quote, Heidegger writes "'*an*' *bedeutet: ich bin gewohnt, vertraut mit, ich pflege etwas.*"[27] It is very interesting for our view of ethics in architecture that although the verb "*wohnen*" usually means "to dwell," in this specific context it collides with the English expression, that "I am accustomed to the place where I reside – to my surroundings."[28] At the same time "*ich pflege etwas*" means that "I am accustomed to do something/ to take care of it/devote myself to it,"[29] in an explicitly ethical notion.

For the Heideggerian philosophy[30] this notion of "in" plays an extremely significant role in the understanding of (human) Being, as a Being that is aware of its existence, a Being for which "to Be" "is an *issue* for it."[31] This is what Heidegger calls Dasein (from the German *Da* = there + *Sein* = Being): the Being that wonders about its existence or that for whom the fact that is there "is an issue." One of the most important ways to understand Dasein, argues Heidegger, is the fact that it exists[32] right there, right now. Dasein is present *in* *in*Humanities: Ethics in architectural *praxis*

the world and for this it is what he calls "being-in-the-world" (German: *in-der-Welt-sein*). Heidegger explains this notion of "in" in two ways: a spatial specific and an existential specific. The first one refers to the definitions of "in" as we saw them before.

> What is meant by "*Being-in*"? Our proximal reaction is to round out this expression to "Being-in 'in the world'", and we are inclined to understand this Being-in as "Being in something" ["Sein in"]. This later term designates the kind of Being which an entity has when it is "in" another one, as the water is "in" the glass.[33]

This relationship can be expanded in a way that *this* desk, there in front of the reader, is in the library; the library is in the university, the university is in the city and so on, until we can finally say that it is in the world. This example, that Heidegger himself gives, reminds us of the Russian dolls that provide an excellent example of a spatial specific notion of "in." Every doll is inside the other, so for each internal doll, its external ones are confined boundaries that have to open up in order for the internals to come into light.

But for Heidegger, the second and most important understanding of "in" is the existential specific. Being-in-the-world means to dwell in this world, to care

Figure 2.3
Being in as spatial-specific (left) and as existential-specific (right). Dwelling in the latter case becomes evident by the presence of the plant by the window that is an allusion to a caring human presence. The human dwells at the threshold.

about it, to be devoted to it and, ultimately, to "have a place on earth." This refers to Dasein's *existential spatiality*,[34] which is fundamental for the understanding of Dasein. According to this view, the human being is not a spiritual entity that is misplaced "into" the space, but on the contrary, it acquires its spiritual existence by the fact that it *is* now, there. This understanding of Being-in-the-world refers to the word "inn" as we already saw.

Here, I would like to argue that this second notion of Being-in-the-world as "inn" shows itself *in* the hyphens. The hyphens have replaced the second "n" of the "inn" and they are needed in order to reconnect the Being with the world. The hyphens emphasise the experience carried by the Being that *is* in the world. The Being and the World are in unity with each other and do not operate only as location-specific entities like the Russian dolls, but coexist, inseparably linked. As Wesley Moriston puts it:

> For Heidegger, man is being-in-the-world, an expression which is *hyphenated* to express the fact that no one of the items constitutive for this phenomenon can be understood in abstraction from the others.[35]

The hyphen as a punctuation mark has been used traditionally in philosophy in order to relate different terms. In a text the hyphen interrupts the always-the-same distance of the common space between the words, offering a bridge between them. It creates a special link between the words, which differs from the single space of the words before and after. But, at the same time, the hyphen keeps them at a distance, it sets them apart or establishes their difference. Postmodern philosophers such as Jean-François Lyotard[36] have placed emphasis on the distance that the hyphen interposes. In these terms the word "Dasein" is different from the dual word "Being-there," since the proximity or the "inn"s of the "Being" in the "there" is not the same. In the word "Dasein" the "Being" is an inherent part of the "there," while the distance in the word "Being-there" is greater. This has to do with the specific possibility that the German language provides for connecting words without the need of hyphens. Nevertheless, Heidegger in the original text writes *in-der-Welt-sein*, choosing to utilise the hyphens and for this we can suggest that the second "n" of the "inn" – the one that draws out attention to ethics – exists *in* the hyphens.

In his "Letter on Humanism," Heidegger responded to Jean Beaufret's question[37] about a possible connection of ethics in the ontology of Being-in-the-world seemingly negatively: "The thinking that inquires into the truth of Being and so defines man's essential abode from Being and toward Being is neither ethics nor ontology."[38] Nevertheless, despite rejecting the question, Heidegger does not reject ethics overall. It is important to understand that for Heidegger, ethics actually means abode or dwelling place: "[t]he word [*ēthos*] names the open region in which man dwells."[39] Obviously this definition of ethics is directly related to his understanding of Being-in-the-world and understanding of "inn" as I have presented here. Furthermore, one could argue that Heidegger does not separate his ontology from ethics: they are one and the same. As Heidegger explains, the various disciplines of philosophy like "logic," "physics" or "ethics" were historically defined during the classical Greek period mainly by Plato's Academy and Aristotle's Lyceum. The "philosophers" before this period were hardly aware of the categorical distinction of these areas of study in philosophy. "Yet their thinking was neither illogical nor immoral."[40] But why is this of any importance to architecture?

Ethics "inn" the world and the praxis of architecture

In the architectural context, when we "do" architecture we constantly make judgements concerning the creation or the demolition of walls. In the extreme cases of the fall of the wall of Berlin and the building of the wall of Jerusalem, these actions are valued as "humane" or "inhuman," expressing as we already saw, an ethical point of view. This evaluation is not a random proposition for the quality of the built environment, but it constantly refers to a wider perspective in life: the world *in* which one dwells. This view sees ethics in a much broader sense than the contemporary professional codes of ethics that constitute the common understanding of ethics in the professions. Every architectural decision pertains to ethics despite the fact that the architect is not constantly aware that he or she "now" operates within a separate field that is distinguished from the activity of doing architecture – similar to the "philosophers" before the classical Greek period that were not immoral before the establishment of ethics as an "official" branch of philosophy. Practicing architecture means to operate with*in* ethics despite the fact that one is not aware of it. It is like the water that the fish swims *in*: no special attention is paid to it, but still it is everywhere one turns. When architects do architecture they make ethical evaluations all the time, since they constantly create and demolish boundaries (analogous to that of Berlin and Jerusalem). These judgements have to do with the thing that architects *do* in order to accomplish their everyday activity of designing and creating space. But what is this activity? What do architects do when they practice architecture?

Inevitably, these questions lead back to the Aristotelian distinction of human activities (*energeiai*) into three categories: *theoria*, *poēsis* and *praxis*. Elsewhere, I have discussed extensively the problems of trying to position architecture as an activity amongst these Aristotelian categories. The most popular dipole is the one that contrasts *theoria* and praxis by assuming that thinking of building a wall is a different thing from doing it. Nevertheless, in the essay *The Switch of Ethics and the Reflective Architect: In-between Practice and Theory*,[41] I consider the operations of praxis and *theoria* as inextricable. *Theoria* is not in any way a set of rules or laws that prescribe practice in advance, but it is participation *in* it. For this praxis is the prerequisite in order for *theoria* to take place. A less popular dipole that hides interesting misunderstandings, especially for architecture, is that which contrasts *praxis* and *poēsis*. Architecture usually is seen as *poēsis*, since it usually has to do with an aim or a goal that stands beyond the mere doing, relating to the production of the physical wall. In the essay *The Flute and the House: Doing the Architecture of Making*,[42] I analysed the examples that Aristotle himself gives (building a house and playing the flute) and concluded by arguing that the poetic aspect of *making* architecture derives from the primary action of simply *doing* it and not the other way around. From both these dipoles, praxis emerges as the mundane bottom ground of architectural activity upon which the other two can flourish.

Nowadays, praxis is usually affiliated with the mere application of abstract ideas, rules and principles preconceived by theory. The building of the wall in Jerusalem, or the fall of the wall in Berlin is usually seen as a mere product of a political thought. Similarly, every wall to be drawn or built in architecture is seen as the outcome of a theory. Here I propose a different comprehension of praxis that is closer to the way that the term was understood in ancient Greece. For the Greeks, the above-mentioned way of acting would have been signified by the term *technē*. Praxis, on the contrary, was an autonomous activity achieved by accomplishing the very action in itself, without seeing it as "application" of some sort and without aiming at a goal that is distinct from the action. The main difference between praxis and *technē* is the fact that praxis involves judgement, exercised by practical reasonableness, that is *phronesis*.[43] Praxis is an action that springs from a tacit understanding that is the outcome of experience and is influenced by a wider world view. In this sense, *ethics inn* praxis becomes responsive to the well-being, not only of the acting self, but also of the others and consecutively towards the whole humanity.

Conclusions

It ought to be a bit clearer now that opposition to "humanism" in no way implies a defense of the inhuman but rather opens other vistas.[44]

Figure 2.4
Beyond the wall

Obviously, the thesis of this paper is not against "humanism" in order to promote barbaric brutality. By problematising the definition of "humanism" as "reasonable faculty," the aim was to reveal the levels of inhumanity that can be achieved by building walls under the excuse of reason. If we remain mere "logical animals," we will always be thinking in oppositional terms, like *in/out*, forgetting the *inn* that dwells in-between. According to these oppositions,[45] when something problematises the human, it is "inhuman," barbaric, bestial or illogical.

It needs more than pure logic to understand architecture as a "humane" activity and even more to transcend it and conceive its divine aspects.[46] However, the way to arrive there is not through pure reflection, but through the close examination of its practice. We have to see in detail how the wall in Jerusalem is built and how the wall in Berlin was demolished, not only as political acts but also as metaphorical architectural practices. Because thinking in architecture is taking place by doing what architects do, i.e. designing, writing or building with their own hands – and at the very same time, reflecting-in-action.

Similarly, the ethics of the architectural practice should be examined *inn* the praxis of actually doing it. And by this I mean, not just by doing what the architects do, but also by dwelling *in* the practice of architecture, by understanding ethics in a perspective of Being-in-the-world and focusing on its hyphens.

This point of view stands against the concept of ethics as an applied discipline. Any proposal to solve the inhuman problem of architecture by merely defining a set of rules (e.g. "Do NOT build walls") that can be applied to its practice as "professional ethics" or "codes of conduct" is condemned to failure. Even the term "applied ethics" is deeply "inhuman" and the notion of "ethics *on* architecture" reveals only hierarchical dominance and empowerment. On the contrary, as presented above, ethics is deeply inherent in architectural activity and for this should be understood through its *inner* relationship to design creativity: ethics dwells in the mundane architectural praxis that is beyond the constant quest for new impressive forms to be published in glossy magazines. Otherwise architecture will keep on building walls without thresholds, and it will keep on being a field for humans to exercise their competence of being inhuman.

Notes

1. Martin Heidegger, "Letter on Humanism," in *Basic Writings from "Being and Time" (1927) to "The Task of Thinking" (1964)*, ed. David F. Krell, London: Routledge, 1993, pp. 213–65, p. 224.
2. Note that by "practice" I don't mean the casual term of "architectural practice" as opposed to "architectural academia," but rather, the activity of generally doing what architects do in their everyday praxis. A more detailed analysis of the term will be provided towards the end of the paper. See also Alisdair MacIntyre's definition of practice in Alasdair MacIntyre, *After Virtue: A Study in Moral Theory*, 2nd edn (corrected, with postscript), London: Duckworth, 1985 (1981), p. 187.
3. As an example of the problematic of defining ethics, see Plato's *Meno* and Witggenstein's *Tractatus Logico Philosophicus*.

4 Maurice Lagueux, "Ethics Versus Aesthetics in Architecture," *The Philosophical Forum*, Vol. XXXV, No. 2, 2004: 117–133, p. 122 [emphasis added].
5 Descartes, "Meditations on the First Philosophy in which the Existence of God and the Distinction between Mind and Body are Demonstrated," in *The Philosophical Works of Descartes*, New York: Dover, 1955, pp. 131–99, p. 145.
6 According to the Oxford English Dictionary there is also a time-specific notion of "in," of "being within limits of periods of time."
in, prep. Oxford English Dictionary online version, 2nd edn, 1989. http://dictionary.oed.com/cgi/entry/50113658?query_type=word&queryword=in&first=1&max_to_show=10&sorttype=alpha&search_id=qd8q-qkX18I-2522&result_place=10 [10.07.2007].
7 *Ibid.*
8 "If we take in our hand any volume; of divinity or school metaphysics, for instance; let us ask, Does it contain any abstract reasoning concerning quantity or number? No. Does it contain any experimental reasoning, concerning matter of fact and existence? No. *Commit it then to flames*: for it can contain nothing but sophistry and illusion." David Hume, *An Enquiry Concerning Human Understanding and Selections from a Treatise of Human Nature, with Hume's Autobiography and a Letter from Adam Smith*, Chicago: The Open Court Publishing, 1912, p. 176 [emphasis added].
9 *in*, prep. Oxford English Dictionary, *op. cit.*
10 Heidegger, "Letter on Humanism," p. 226.
11 *in*, prep. Oxford English Dictionary, *op. cit.*
12 *Ibid.*
13 *Ibid.*
14 *in-*, prefix, Oxford English Dictionary online version, 2nd edn, 1989. http://dictionary.oed.com/cgi/entry/50113661?query_type=word&queryword=in&first=1&max_to_show=10&sort_type=alpha&search_id=qd8q-qkX18I-2522&result_place=10 [10.07.2007].
15 *in-*, prefix Oxford English Dictionary online version, 2nd edn, 1989. http://dictionary.oed.com/cgi/entry/50113663?query_type=word&queryword=in&first=1&max_to_show=10&sort_type=alpha&search_id=qd8q-qkX18I-2522&result_place=10 [10.07.2007].
16 *inhuman*, a. (n.), Oxford English Dictionary online version, 2nd edn, 1989. http://dictionary.oed.com/cgi/entry/50117015?query_type=word&queryword=inhuman&first=1&max_to_show=10&single=1&sort_type=alpha [10.07.2007].
17 *Ibid.*
18 Heidegger, "Letter on Humanism," p. 227, see also pp. 228–29.
19 in, adv. Oxford English Dictionary online version, 2nd edn, 1989. http://dictionary.oed.com/cgi/entry/50113657?query_type=word&queryword=in&first=1&max_to_show=10&sort_type=alpha&search_id=VBow-Gm8570-5558&result_place=10 [11.07.2007].
20 *in*, v., Oxford English Dictionary online version, 2nd edn, 1989. http://dictionary.oed.com/cgi/entry/50113656?query_type=word&queryword=in&first=1&max_to_show=10&sort_type=alpha&search_id=VBow-Gm8570-5558&result_place=10 [11.07.2007].
21 *Ibid.*
22 *inn*, v., Oxford English Dictionary online version, 2nd edn, 1989. http://dictionary.oed.com/cgi/entry/50117304?query_type=word&queryword=in&first=1&max_to_show=10&sort_type=alpha&result_place=10 [11.07.2007].
23 inn, n. Oxford English Dictionary online version, 2nd edn, 1989. http://dictionary.oed.com/cgi/entry/50117303?query_type=word&queryword=in&first=1&max_to_show=10&sort_type=alpha&result_place=10 [10.07.2007].
24 *Ibid.* [emphasis added].
25 Martin Heidegger, *Being and Time*, trans. by John Macquarrie and Edward Robinson, Oxford and Cambridge, MA: Blackwell, 1962/1927, p. 80.
26 *Ibid.*
27 *Ibid.*
28 *Ibid.*
29 *Ibid.*
30 The reader should be aware that any attempt to reduce the philosophy of Heidegger into a few quotes is a doomed attempt. At the same time, though, it is impossible to understand my argu-

ments that follow, without having even a vague idea of the core concepts of Heidegger mentioned here. This is a warning for the reader and a reminder for the author of the inherent impossibility of his task.

31 Heidegger, *Being and Time*, H 12, p. 32 [emphasis in the original].
32 *Ibid.*, H 53, p. 78.
33 *Ibid.*, H 53–54, p. 79 [emphasis in the original].
34 *Ibid.*, H 56, p. 83.
35 Wesley Morriston, "Heidegger on the World," *Man and World*, 1972: 5(4), pp. 452–467, p. 265 [emphasis added].
36 See, for example, Jean François Lyotard, *The Hyphen: Between Judaism and Christianity*, New York: Amherst, 1999.
37 Beaufret says, "What I have been trying to do for a long time now is to determine precisely the relation of ontology to a possible ethics." Heidegger, "Letter on Humanism," p. 255.
38 *Ibid.*, p. 259.
39 *Ibid.*, p. 256.
40 *Ibid.*
41 Leonidas Koutsoumpos, "The Switch of Ethics and the Reflective Architect: In-between Practice and Theory," in *Reflections on Creativity: Exploring the Role of Theory in Creative Practices*, Hamid Van Koten (ed.), Dundee: University of Dundee, 2007.
42 Leonidas Koutsoumpos, "The Flute and the House; Doing the Architecture of Making," in *The Politics of Making*, Mark Swenarton, Igea Troiani, and Helena Webster (eds.), *Critiques: Critical Studies in Architectural Humanities*, London: Routledge, 2007, pp. 105–116.
43 Richard Coyne and Adrian Snodgrass, *Interpretation in Architecture: Design as a Way of Thinking*, London and New York: Routledge, 2006, p. 112.
44 Heidegger, "Letter on Humanism," p. 250.
45 *Ibid.*, p. 249.
46 Arguably, for Heidegger, the human Being operates in-between its animal appearance and its divine potential.

Chapter 3

Cultivating architects

History in architectural education

Alexandra Stara

Introduction

What is the relevance of history in architectural education today? The majority of aspiring architects beginning their studies seem pretty clear about how being taught technology or management will assist with the development of their design projects and, eventually, their professional skills – but the role of history remains elusive. This may have something to do with the kind of history students are presented with, but perhaps also with the resistance of the humanities to be translated to mere "skills". This chapter will revisit history as a non-instrumental mode of enquiry and will argue that such a historical understanding of architecture is inseparable from its proper understanding as a creative act.

Theory

We cannot make or build without thinking. Ideas are intrinsic to architecture. But what such ideas might consist of and where they might legitimately come from has become a complex affair in our late modern world. Despite gestures towards a balancing of technological and humanist elements in the education of architects, the real issues remain largely ignored. What is at stake is not a matter of disciplinary pluralism or "interdisciplinarity" – another catchphrase of our times – but instead the very nature of architecture as a creative process and the ways in which it can be meaningfully unravelled and taught. The disciplinary divide, which radically separates word, number and image, has acute repercussions in architecture, perhaps the most synthetic and situated of cultural phenomena. This separation is especially problematic for education, as the false security of architectural representation and its "make-believe" world is shielded from the complexities of the real.

Among other things, postmodern critique attacked the technological bias of modernism through a turn to cultural discourse as an essential reference for architecture. For several decades now, architectural practice has been increasingly dependent on various theories and critiques, producing work that certainly cannot be accused of lack of ideological content. Even more so, the teaching of architecture has become highly dependent on theory, which has overtaken history as the discourse of choice for supporting architectural education. The intention of theory to offer a more productive and relevant mode of enquiry than the stale incantation of history in traditional "beaux-arts" style education may be commendable. However, the question is to what extent this enquiry can have results relevant outside the limits of each theoretical framework.

Theory conceptualises and proposes to shed light on its subjects through varying degrees of reduction and measuring against specific terms of reference. According to Hans-Georg Gadamer, theory "is a tool of construction by means of which we gather experiences together in a unified way and make it possible to dominate them".[1] Theory as a construct implies that one theory will follow another and, thus, each theory can only ever have conditional validity. Furthermore, operating as the opposite of practice, theory has an almost "privative" character. Gadamer writes, "something is meant only theoretically when it does not have the definitively binding quality of a goal or action".[2] There is a fundamental difference between a conceptual endeavour which attempts to influence practice as a closed system external to it, and a mode of reflection which informs and is informed by practice through an open dialectic engagement with it. It is only the latter which can assist with the redressing of the technological bias of architecture and the reinstatement of its relevance in the continuity of culture as a primary creative endeavour.

Common ground

The modern dominance of instrumental knowledge and the accompanying reduction of all non-technological matters to ephemera has had profound consequences in areas of culture beyond the direct involvement of technology. One such is the emancipation of the arts from the common ground of culture and their fragmentation into various domains of self-reference. Dalibor Vesely argues that architecture has become "a mosaic of expert knowledge brought together either as abstract systems or as the intuitive improvisations of personal vision".[3] In order to escape the conditional and arbitrary and to be able to fulfil its role as embodiment and foundation of culture, architecture needs to be understood in terms of participation in and emergence from a shared ground. If it is to be more than an expression of personal tastes or a mere commodity, if it is to be convincing as a response to cultural conditions and aspirations, then, as a creative process as well as a built environment, it needs to be grounded in the continuity of place and shared identity.

Karsten Harries reminds us that our dwelling is always a dwelling with others, so the problem of architecture is inevitably also the problem of community.[4] Community will, as it always has, define itself through fundamental acts of participation where identity and order are celebrated and confirmed. The highest role of architecture remains what it has always been: to invite these participatory acts, these festivals where individuals come together to affirm themselves as part of the community – ultimately, to celebrate "those central aspects of our life that maintain and give meaning to existence".[5] The notion of community might appear problematic at a time when individuality and difference are promoted as the highest achievements on almost every level of human endeavour. However, the common realm implied here is not a wilful collective of sorts, but the very ground whereupon culture is made possible as a fundamental aspect of our humanity. In that sense, shared identity is formed and made manifest as a latent, open structure, through a reciprocity between the same and the different, the actual and the possible, perpetually in a state of becoming – what one might call a dynamic process, rather than a conclusive one.

Recognising the common ground of culture, always latent and even partly obscured, is nevertheless a condition for the possibility of meaningful, relevant creativity. At the beginning of his *Truth and Method*, Gadamer revisits Giambattista Vico's notion of *sensus communis*, as part of his juxtaposition of humanist tradition and modern scientific method. According to Vico, writes Gadamer, "what gives the human will its direction is not the abstract universality of reason but the concrete universality represented by the community of a group, a people, a nation or the whole human race".[6] Developing the *sensus communis* is of decisive importance for living and the most important thing in education.[7] This, however, cannot happen through emancipated, theoretical knowledge, which engages the faculties of reason and abstraction and cannot fully exhaust the sphere of the knowable. *Sensus communis*, as a sense "acquired through living in the community and determined by its structures and aims",[8] is engaged within an altogether different mode of knowing, that of the humanities. "For their object, the moral and historical existence of humanity, as it takes shape in our words and deeds, is itself decisively determined by the *sensus communis*."[9]

History

In other words, because of its very nature, we cannot understand the ground of culture properly, unless we understand it historically. For architecture, this becomes a matter of immediate responsibility, as this understanding (or absence thereof) has direct implications on the fabric of our everyday lives. The problem of much architectural history in education, however, is that it has largely remained an objectifying project, in sole pursuit of fact and intolerant to speculation. This kind of self-referential history is what the proliferation of theoretical discourses has attempted to address, only to fall into the same trap of abstraction and loss of relevance. Mapping the past and its relationship to the built environment as an evolution of form and technique performs a drastic reduction on architecture and thus separates itself from the creative process. It is a cipher, which may be unlocked with the right keys – for example, learning the names, forms and proportions of the five classical orders – but cannot be translated into practice, unless one still believes nineteenth-century historicism is the way forward. If the movement of history has something more significant to teach us, then it cannot be properly understood as the object of a science of history. Such a history is disciplinarily isolated and can only make sense for its own sake. Like an archaeology of building, it remains focused on aesthetics, techniques and functions, as autonomous, objectified units.

Architecture both shapes and is shaped by the specificities of place and time, of a particular culture and its conditions. At the same time, it reflects universal themes such as dwelling, shelter and ritual, and is a crucial mediator in fundamental polarities of the human condition such as nature/artifice, embodiment/ideal, public/private and so on. A history addressing architecture as such a phenomenon is one that does not reduce its subjects to mere objects but allows for the whole spectrum of culture to be potentially filtered through them. It is a history that reads architecture as a deeply situated cultural phenomenon while recognising the universality of its themes, allowing it to emerge as a reciprocal relationship between the multiple and latent strata that contribute to its formation.

As recognised by the long tradition of the humanities, such a history is a source of truth very different from theoretical reason. It proceeds based on concrete examples, building an open enquiry through their interpretation, as a movement between the specific and the general, the known and the unknown. The way understanding occurs in such a hermeneutic enquiry is not a methodological concept but an active engagement. This understanding "implies the general possibility of interpreting, of seeing connections, of drawing conclusions". Crucially, through its balanced movement between the actual and possible "it constitutes a state of new intellectual freedom".[10] That intellectual freedom and creativity are not an unbridled foray into the unknown, but a dialogue between what is already here and what can follow, may be challenging for our modern sensibilities. However, despite their dominance in the realm of the arts, the idea of an emancipated creation and the cult of originality are concepts invented in the realm of technological thinking, offering a reduced and rather skewed version of those areas of human endeavour that positive science simply cannot come to grips with.

The freedom allowed for in this dialectical understanding of history is akin to the controlled workings of imagination in the mode of analogy. Anathema to the scientific understanding of history, imagination in historical enquiry has been rehabilitated in the context of postmodern discourse by such figures as Hayden White.[11] But it is Paul Ricoeur's reading of the analogue as the dialectic between "same" and "other" that strikes as most convincing, offering up a space of imagination and creative freedom, while preserving the important distinction between history and fiction.[12] Simultaneously accepting the irrevocability of the past – hence "other" – and the survival of its traces which partly constitute the present – hence "same", Ricoeur proposes analogy as the medium for beginning to resolve this tension. He writes:

> The past is indeed what is to be re-enacted in the mode of the identical. But it is so only to the extent that it is also what is absent from all our constructions. The analogue, precisely, holds within it the force of re-enactment and of distancing, to the extent that *being as* is both being and not being.[13]

History, then, is not a bridge attempting to span a divide between past and present, because the past "is actually the supportive ground of the course of events in which the present is rooted". The real focus of history is not an unattainable objectivity, nor an elusive compensation for temporal distance, but recognition of that distance as the space of understanding, "filled with the continuity of custom and tradition".[14]

Cultivation

The title of this chapter refers to the education of architects as cultivation, while proposing history as an essential component thereof. The notion of cultivation has a long trajectory in the humanities and helps bring together the key themes discussed here. Martin Heidegger offers the definitive challenge to an instrumental understanding of building by linking it ontologically to dwelling.[15] In the course of his discussion, he shows that inherent in the notion of building are ideas of cherishing, protection, preservation and care – that is, of cultivation, in its original reference to working with the land. At the same time, building is constructive – an aspect of its meaning which is carried through the word "edifice", from the Latin *aedificare*.[16] While the caring for the land and raising up of edifices are different modes of building, they are not inherently conflicting (although, of course, we know that they can be, and are frequently so). The complimentary nature of this relationship is brought forward in the meaning of cultivation, which, to this day, carries within it the reconciliation of nature and artifice; but also intriguingly in the meaning of edification, which although stands for the more constructive aspect of building, has come to be a near synonym for cultivation in its modern sense, as intellectual and moral improvement.

This sense of cultivation is inherited from the early nineteenth-century notion of *Bildung*, broadly understood as an attempt to shape modernity through a reinvestment in humanism. Although there are considerable pitfalls in this notion and its precise philosophical articulation, it remains eminently useful as a vehicle for rethinking the relationship between individual to culture and the role of the humanities in this dialogue. Developing from a close association with the idea of cultural formation – designating the proper way of shaping one's natural attributes in accordance with the cultural context – *Bildung* eventually becomes something "both higher and more inward, namely the disposition of mind which, from the knowledge and the feeling of the total intellectual and moral endeavour, flows harmoniously into sensibility and character".[17] This notion of *Bildung* rests on the recognition of the universal dimension of humanity, which is accessed not as a theoretically detached idea, but in reciprocity with one's own situation. This movement away from oneself is necessary for the understanding of the other, which is not a Cartesian object complete in and of itself, but shares with the self a ground that allows (re)cognition. A return to oneself is an essential part of this movement – so not an abandonment of the particular for the universal, but its opening up and enrichment.

Crucially, cultivation as *Bildung* comprises a strong ethical dimension. The aspiration to move beyond the boundaries of one's own particularity is posited as no less than a duty, but also the sheer ability to understand in this dialectical manner requires an engagement which cannot be dissociated from moral considerations. To know as part of cultivation is to assume a stance, a thesis – which means position in the original Greek – and is therefore ineluctably woven with judgement and responsibility. This resonates with Heidegger's positing of preservation and care as essential dimensions of building. In the radically shifting culture of the late Enlightenment, the project of *Bildung* was an attempt to establish continuities and secure a ground for which to care and on which to build. Although in an advanced modernity now, it is no less difficult to know what to care for and how. An engagement with history as historical grounding, as the fundamental historicity of culture and of the human condition as a whole, is an essential base for the possibility of finding appropriate answers.

Creativity

It is clear that such an understanding of history requires an active engagement from its students, not unlike the process of design, which it is intended to support. It is part of the project of understanding architecture as a cultural phenomenon in order to be able to make it, but also understanding what this "making" consists of.

There is deep reciprocity between freedom and contingency in the design process, which is mirrored in the study of history. The space of creativity in the making of architecture emerges out of extended negotiations between desire and possibility, on one hand, and what is already there, as nature and culture, physically and intellectually, on the other. The idea of a *tabula rasa* is nowhere more misconceived than in architecture, where even the most extreme of sites belongs to the actuality and concreteness of the world, alongside all the binding parameters of the project's commission. In stark contrast to the image of a blank sheet awaiting the first markings of the architect–creator, each architectural situation presents an immensely complex and varied structure of givens requiring translation into a unified medium. It is in this act of translation – for which, as we know from poetry, there is never a single right answer – that creative freedom is made possible.

The creative dimension of interpretation and, conversely, the interpretative aspect of creativity, establish the relationship between architectural history and design as an essential one. The questions of how we should approach design, how we should learn to make decisions, to favour certain choices over others, cannot be answered through design-specific tools alone. History in architectural education cannot be an interval in the curriculum, a respite from the serious business of design or, as for many, an unwelcome distraction by discourse into an otherwise "visual" discipline. The ground upon which architecture was, is and will be made, the ways in which this can and should/should not happen and the measure of its success or failure are all issues woven with architecture's historicity. None of this, of course, relates to an encyclopedic history, where one might look for direct answers as if for entries in a dictionary. What is at stake, as briefly outlined in this chapter, is first and foremost an attitude, a stance towards historical understanding which appreciates the nature of architecture as both a manifestation and foundation of culture.

Arguably, this is the highest challenge for architectural education as a whole – not to provide the answers as ready-mades, but to teach the means through which one can begin to ask the appropriate questions and negotiate ways of answering them. History, therefore, is an essential medium and a way into architecture. If it is to be useful, the disciplinarity of history needs to be transcended. It should dissolve in the process of articulating and manifesting the nature of architecture as a situated cultural phenomenon. Understanding history as a constituting dimension of architecture brings to the fore our own historicity. We are also situated in history and conditioned by it. As Gadamer tells us, historical understanding is ultimately a self-understanding. We cannot understand ourselves and the way to be, let alone to design, unless we understand our own historicity.[18]

By way of conclusion, an often-quoted passage from Aristotle's *Metaphysics*, which is once again relevant here. "We consider that the master-craftsmen [*architektones*] in every profession are more estimable and know more and are wiser than the artisans [*cheirotektones*] because they know the reasons of the things which are done."[19] It is not a matter of technique or knowledge of "what" that distinguishes the masters or "first" makers from the others, it is the "why". What we're dealing with here is responsibility rather than mere ability, understanding rather than mere knowledge. The importance of history for learning and making architecture is inextricably woven with architecture's ethical function, and the fulfilment of its highest claim to ongoing cultural relevance. In an educational context of increasing specialisation, where perpetual innovation is promoted as an unqualified virtue, and where, ultimately, procedural concerns and commercialisation supersede academic purpose, it is imperative that such aspirations are not abandoned.

Notes

1 Hans-Georg Gadamer, *Truth and Method*, J. Weinsheimer and D.G. Marshall (trans.), second revised edition, London: Sheed & Ward, 1989, p. 454.
2 *Ibid.*
3 Dalibor Vesely, *Architecture in the Age of Divided Representation*, Cambridge, MA: MIT Press, 2004, p. 3.
4 Karsten Harries, *The Ethical Function of Architecture*, Cambridge, MA: MIT Press, 1997, p. 13.
5 *Ibid.*, p. 365.
6 Gadamer, *Truth and Method*, p. 21.
7 *Ibid.*, p. 20.
8 *Ibid.*, p. 22.
9 *Ibid.*, pp. 22–23.
10 *Ibid.*, p. 260.
11 See, for example, his *Metahistory: Historical Imagination in Nineteenth Century Europe*, Baltimore, MD: John Hopkins University Press, 1975.
12 Paul Ricoeur, *The Reality of the Historical Past*, Milwaukee, WI: Marquette University Press, 1984.
13 *Ibid.,* p. 36.
14 Gadamer, *Truth and Method*, p. 297.
15 Martin Heidegger, "Building Dwelling Thinking", in *Poetry, Language, Thought*, A. Hofstadter (trans.), New York: Harper & Row, 1971, pp. 145–61.
16 *Ibid.*, p. 147.
17 Wilhelm von Humboldt, *Gesammelte Schriften*, Akamemie ed., Vol. VII, Part I, p. 30, as cited in Gadamer, *Truth and Method*, pp. 10–11.
18 Gadamer, *Truth and Method*, p. 260.
19 Aristotle, *Metaphysics*, 981a30–b2, as cited in Alberto Pérez-Gómez & Louise Pelletier, *Architectural Representation and the Perspective Hinge*, Cambridge MA: MIT Press, 1997.

Chapter 4

The architect as humanist

Leslie Kavanaugh

Architecture has historically been a profession derived from two traditions: the master builder and the humanist. The one is about "technique;" the other about being an educated person in society at large. The one is learned from the older, more experienced professionals; the other is more textual and is contemplated, yet also ideally guided by wise teachers. The one is "practiced;" the other is "theorized." Certainly, today, the technique of the master builder is "fast" – with technologies of construction, representation and material development progressing at such a pace as to be almost ungraspable. In the humanist tradition the pace is "slow" – with thought that is not facile, and with ideas that take time to mature and to work themselves into the soil so to speak. Following Aristotle, the humanist tradition exalted man as an animal that reasons, and most specifically an animal that reasons for *himself*. These two traditions have often been at odds with each other, yet this state of affairs is not a necessity.

The humanist tradition is characterized most decidedly by an increased freedom of the individual – economically, politically, and most importantly intellectually. Human thought became increasingly a matter of reason, in place of conformity to authority. One of the most succinct descriptions of this humanist turn is in the 1784 essay from Immanuel Kant, written for the popular press, entitled: "An Answer to the Question: What is Enlightenment?"[1] Kant does not mince words. In his opening sentence, he boldly states that enlightenment is man's escape from self-imposed childishness.[2] Only a child would submit unquestioningly to the opinion of others. A man – a humanist – reasons for himself. Yet, in reasoning, a man must take responsibility for his own enlightenment, and this resolve to escape immaturity takes courage. Even though man increasingly had the freedom to think, out of sheer laziness, it was easier not to think. "Why think," Kant ironically asks, "when you can pay?" – much easier to let someone else think for you.

So, in one sense, the "humanities" as a scholarly discipline, could be seen to arise from the Enlightenment tradition, from the challenge to think for oneself, from the imperative to escape from self-imposed ignorance. Yet taking on the mantle of this responsibility, one is immediately confronted with the questions: What is thinking? What to think? How to go about thinking?

This question about knowledge is as old as human history, as old as the oldest recorded attempts at thinking, that is to say, the ancient Greeks. Plato, for example, asked the question of true knowledge in the dialogue entitled *Theaetetus*, and unable to come to a definitive answer, continued the inquiry in the dialogue entitled the *Sophist*. Can a thinker know everything, Plato asks? Indeed, can we know what even a single concrete individual thing is? In the case of the dialogue the *Sophist*, Plato asks "What is a 'sophist'?" as a particular kind of thinker in ancient Greece, and broadly speaking Plato wishes to ascertain whether the sophist as opposed to the true philosopher, has true knowledge. Most importantly for our discussion here, is the distinction that Plato makes in this dialogue about *sophos technē*, or the ability to know wisdom. Wisdom is *sophos* in Greek, a word that of course forms the root of the word philosophy – *philos + sophos*, or the love of wisdom. The Greek word *technē*, we naturally recognize as the root of our English word "technology." Yet in ancient Greece, most specifically in the dialogue the *Sophist*, Plato does not make the kind of determinations that we make in the contemporary meaning of this word; that is to say, that technology deals with applied science. Rather, *technē* is the *art*, or the *ability to know* generally speaking.

Plato divides this art of knowing, *technē*, into the acquired and the productive knowledge. The knowledge that is acquired can be done so either voluntarily or forcefully. Productive knowledge can be attained through the production of likenesses or semblances. For Plato, any knowledge needs to conform to the Ideal in order to constitute true knowledge or wisdom. Yet, most importantly for our discussion, a distinction is made as to the true and the false, as opposed to the "art" and the "science" of *technē,* or the *art of the ability to know*. Putting "art" and "science" as oppositional categories was a much later project of modernism, which is to say, the sixteenth and seventeenth centuries. The fact that the architect as humanist, or technician, or master builder became divisions within the profession of architecture is also of recent invention. Yet in ancient Greece, *technē* is the art, or the ability to know.

Martin Heidegger also acknowledged this important word *technē* in its originary sense in his essay "*The Question Concerning Technology*." Heidegger interprets technology as a mode of revealing truth, *alétheia*, not as mere instrumentality or utility of technique. "Technology," he states, "is no mere means. Technology is a way of revealing."[3] Unfortunately, technology has been reduced to mere technique, to a highly "framed" manner of knowing that decides beforehand what is worth knowing, and its possible utility, and also perhaps its projected profitability. As such, *technē* is not thought in its originary sense, and indeed not thought in its potential to reveal the truth. "There was a time," Heidegger argues, "when it was not technology alone that bore the name *technē*. Once that revealing that brings forth truth … was also called *technē* …; art was simply called *technē*."[4] Art, or *technē*, is the ability to know in its acquisitive and productive sense. For Heidegger, "so long as we represent technology as an instrument, we remain held fast in the will to master it. We press on past the essence of technology."[5]

In thinking *technē* in its originary sense, *technē* is the art, or the ability to know; yet this thinking is accomplished by humans, in their full humanness. "Thinking accomplishes the relation of Being to the essence of man."[6] Yet, Heidegger goes on to say, man must accomplish thinking free of the "technical interpretation of thinking,"[7] without the framing of knowledge as instrument, or making, or practice. Indeed, Aristotle defined man as the beast who reasons: the *animal rationale* (*zoon logon echon*). Yet, for Heidegger, this definition of man thinks of man on the basis of his animality, not his humanness. "Humanism is opposed because it does not set the *humanitas* of man high enough."[8] Humanism as defined by the Western tradition still does not think of man in his true dignity. Not only is man the animal who reasons, but he is the animal who is reduced to the mere rational, reduced to a being who is not defined also by "care," by compassion, by empathy. These definitions also characterize what it means to be human. Thus Heidegger concludes, "this is humanism: meditating and caring, that man be human and not inhumane, 'inhumane,' that is, outside his essence."[9]

In his *Letter on Humanism*[10] published in 1947, Heidegger analyses fragment 119 from Heraclitus: *ethos anthropoi daimon*, which he translates as, "Man dwells, insofar as he is man, in the nearness of god."[11] In this short passage Heidegger links Being with dwelling and thinking with the essence of being human. The letter, as such, was a response to the French philosopher Jean Beaufort, in part to question the notion of "humanism," both as an intellectual legacy and as a possible ethic after the devastation of the Second World War. Heidegger attempts to enlighten this relatedness between man and Being with the word *ethos*, with fragment 119 from Heraclitus: *ethos anthropoi daimon*, normally translated as, "a man's character is his spirit." In other words, the inborn nature of man becomes his guiding spirit, his god. To take the fragment still further apart: *ethos* means character; *anthropoi* means man; and *daimon* means gods, a supernatural intelligence, a guardian spirit, a genius, or a person of great skill or energy. Yet, according to Heidegger's interpretation, "this translation thinks in a modern way, not a Greek one." Instead, Heidegger translates the Greek *ethos* as "abode, dwelling place." For him, *ethos*

> names the open region in which man dwells. The open region of his abode allows what pertains to man's essence, and what is thus arriving, reside in nearness to him, to appear. The abode of man contains and preserves the advent of what belongs to man in his essence. According to Heraclitus' phrase this is *daimon*, the god. The fragment says: Man dwells, insofar as he is man, in the nearness of god.[12]

Yet, would this Being that is near help define man as "humane" as opposed to "inhumane?" The proximity of Being to man does not necessarily make him more humane. Man must make an open space for the presencing of Being. Man is the guardian of Being for Heidegger. Yet fundamentally, dwelling is the nature of man in the nearness of god. The *ethos*/abode of man is what dwells nearest to him. Where a man dwells and has his abode, there also dwells his god, nearest.

What, then, is the essence of the man who thinks humanely? In this fragment of Heraclitus,[13] Heidegger sets the scene. A group of tourists, ever hungry for novelty, come in search of the great philosopher Heraclitus. What, pray tell, does a philosopher look like? Shall they catch him in the act, in the very process of philosophizing, of thinking? They no doubt expect something grand, extraordinary, life transforming; for this is a great man, of worldwide fame, yes? They expect to touch his robes and be transformed in some dramatic way, without of course undertaking the labor of thinking themselves. However, they are disappointed. Heraclitus looks like an ordinary man – even common – ugly perhaps. His "appearance" does not correspond to the rumored greatness of the man. Perhaps he is slight of stature, old, and a bit shrunken by the strain of thinking. Perhaps he has not seen a generous meal in quite a long time. He has known poverty, he has known failure; he has known destitution. To philosophize is to be met with more failure than success. Consequently, most profoundly, he has known aloneness, the isolation of not being understood. He humbly warms his hands at his meager fire; he is about his habitual ablutions. They do not see anything special about either the place or the man. Instead, what they see is familiar, habitual, and even commonplace. The human-all-too-human Heraclitus inhabits an abode that appears everyday; in short, anything but extraordinary.

The group, after peering in through the low portal, dare not cross the threshold. They turn away slightly angry that they have come so far for nothing – come so far for what is so common and habitual, come so far for what was always so near. They feel cheated out of a sublime experience. "In this altogether everyday place he betrays the whole poverty of his life,"[14] Heidegger relates. They withdraw disheartened. Yet Heraclitus catches a final glimpse of them at the door, and reassures them by saying: "Here too the gods are present."

> *Kai entautha*, "even here," at the stove, in that ordinary place where every thing and every condition, each deed and thought is intimate and commonplace, that is, familiar [*geheuer*], "even there" in the sphere of the familiar, *einai theous*, it is the case that "the gods are present." Heraclitus himself says, *ethos anthropoi daimon*, The (familiar) abode is for man the open region for the presencing of god (the unfamiliar one).[15]

Nonetheless, merely "even here," the commonplace and familiar might rather be translated "especially here." "Especially here" in the realm of the familiar, the gods dwell near to man. Of course, quite obviously the word "familiar" is etymologically linked with family, or *oikos* in Greek, the familial household. As in "Building Dwelling Thinking," habitation is again linked to the habitual, the common, to the familiar. Being habitually dwells alongside what is familiar. So "especially here," the gods dwell for man in the familiar commonplace, the place that is nearest. "[M]an occurs essentially in such a way that he is the 'there' [*das "Da"*], that is, the lighting [opening, clearing] of Being."[16] "Dwelling is the essence of 'being-in-the-world.'"[17] In this "average everydayness," in these concrete structures of human beings in their predominant state going about their daily tasks, Heidegger offers an existential analysis of Dasein: "Dasein is in every case concernful Being-alongside."[18] This Being-alongside is *Sein bei* in German – *bei* meaning "at the home of," "*chez*"; i.e. Dasein is the care or solicitude of being at the home *of* Being, at home *with* Being. Unfortunately, this care,

> this Being-with-one-another dissolves one's own Dasein completely into the kind of Being of "the Others," in such a way, indeed, that the Others, as distinguishable and explicit, vanish more and more ….We [philosophers] take pleasure and enjoy ourselves as "*they*" [*das Man*] take pleasure; we read, see, and judge about literature and art as *they* see and judge; likewise we shrink back from the "great mass" as *they* shrink back; we find "shocking" what *they* find shocking.[19]

So the philosopher is the *anthropos* of the *ethos antropoi daimon* like any other *anthropos*, any other "human," yet something more than the commonplace is expected of the philosopher – something more humane from the human-all-too-human Heidegger. He, too, would come "to betray the whole poverty." Yet for the moment, Heidegger forestalls with the warning:

> But if man is to find his way once again into the nearness of Being he must first learn to exist in the nameless. In the same way he must recognize the seductions of the public realm as well as the impotence of the private. Before he speaks man must first let himself be claimed again by Being, taking the risk that under this claim he will seldom have much to say. Only thus will the preciousness of its essence be once more bestowed upon the word, and upon man a home for dwelling in the truth of Being.[20]

In order to escape the seductions of the public realm – fame, diffusion of energies, compromises in the name of expediency, losing the voice of his soul in the clamor and chatter of men – Heraclitus retires to his modest fire. Yet the dilettante intellectuals have come to him expecting greatness. In the private realm of his home, Heraclitus dwells with the gods, yet man sees only the appearances of a man altogether too familiar with the familiar, with destruction and destitution. He dwells with the gods in his *ethos*, his abode, as the guardian of Being, "thinking, belonging to Being, listen[ing] to Being."[21] Heraclitus is "at home" (*Sein bei*) in his simple intimacy with Being, in his belongingness (*Gehörigkeit*), dwelling with his god, nearest. The gods gather around the hearth, the meager everyday fire that warms the hands of Heraclitus. In the habitation of man, the gods dwell near in the habitual. Inhabitation reveals an ease with the things about us, a certain belonging to or having (*habere*), not in the sense of possession, but having a familiarity with, a related intimacy with our surroundings. To dwell-in is the Latin *in* = *habitare*. We dwell familiarly with things.

Heidegger was *not* a humanist, rather an anti-humanist. Heidegger thought that Aristotle's definition of man as the rational animal did not set the bar high enough, did not answer the question of what was "humane" in the "human." Nevertheless, the current hegemony of the image, indeed as Baudrillard has framed it, where the image subsumes or replaces the object, has abandoned both the technological and the humanist traditions in architecture. Architecture no longer "gets its hands dirty" with the materiality of building; rather has the tendency to skate upon the shimmering superficiality of *simulacra* – collections of visual sensations passing rapidly before the eye, space flattened out into mere two-dimensions, devoid of engagement.

Following Plato's *Sophist*, where *technē* is "*to be able to know*" – perceiving, conceptualizing, materializing, making – knowledge is multiple. What knowledge is *not*, is information. So, perhaps the problem in architecture in not necessarily to find a way back to the "word" from the hegemony of the "image"; rather to let architecture "unfold essentially" in all her manifest ways, yet always remembering that architecture is about "space" and not image; and that this space is "humane." Word and image are not juxtapositions, rather alternative ways of *being able to know*, possible *technē*. If Immanuel Kant challenges us to take up the responsibility for thinking, to shake off our "childishness," our laziness, in order to become enlightened rational beings, Heidegger goes further. The thinker should not seek the expedient, the hurried, the excited; rather, the humane dwells in the essence of man, in his thinking.

To be a humanist is to think, and to think most importantly for oneself, with care, humanely. If we follow Aristotle in defining man as the animal who reasons, might we say, then, that to think is perhaps reduced to the instrumentality of technology though not in its originary sense. To be a humanist is to care, to abide in thought. The architect abides by the inquiry into the *technē*, the modes of "being able to know." This abiding need not be dichotomous, conflictual, or contentious. The architect abides, humanely, in order to dwell humanly.

The architect as humanist takes up the challenge to think for oneself, to escape from easy concepts, to be responsible for "knowing." So perhaps the question is not: "What to think?" Rather the imperative, following Kant, "Dare to know!"

Notes

1 Immanuel Kant, "Beantworung der Frage: Was ist Aufklärung?", *Berlinische Monatsschrift* 1784: 4, pp. 481–94. For a survey of the development and influence of Enlightenment ideals, see, Louis Dupré, *Enlightenment and the Intellectual Foundations of Modernism*. New Haven, CT: Yale University Press, 2005.
2 The term that Kant uses here is "*Unmündigkeit*," which is a legal term indicating the inability to represent oneself in court or more broadly speaking the *in*ability to be legally held accountable for one's actions; that is to say, children, women, and madmen.

3 Martin Heidegger, "The Question Concerning Technology" in *The Question Concerning Technology and Other Essays,* translated and introduced by William Lovitt, New York: Harper & Row, 1977, p. 12.
4 Heidegger, "The Question Concerning Technology," *op. cit.*, p. 34.
5 *Ibid.*, p. 32.
6 Martin Heidegger, (1977) "Letter on Humanism," translation of *Brief über den Humanismus* by Frank A. Capuzzi with J. Glenn Gray, published in the volume, David Farrell Krell (*ed.*), *Martin Heidegger: Basic Writings*, San Francisco, CA: Harper & Row, p. 193.
7 Heidegger, "Letter on Humanism," *op. cit.*, p. 194.
8 *Ibid.*, p. 210.
9 *Ibid.*, p. 200.
10 *Ibid.*, pp. 193–42.
11 *Ibid.*, p. 233.
12 *Ibid.*
13 Derived from Aristotle, *De parte animalium, I,* 5, 645a 17.
14 Heidegger, "Letter on Humanism," *op. cit.*, p. 234.
15 *Ibid.*, p. 234.
16 *Ibid.*, p. 205.
17 *Ibid.*, p. 236.
18 Martin Heidegger, *Being and Time*, John Macquarrie and Edward Robinson (trans.), San Francisco, CA: Harper & Collins, 1962, p. 180.
19 *Ibid.*, p. 164.
20 Heidegger, "Letter on Humanism," *op. cit.*, p. 199.
21 *Ibid.*, p. 196.

Chapter 5

Migration, emancipation and architecture

James McQuillan

> You shall be a fugitive and a wanderer over the earth.
>
> Gen. 4:12.

Introduction

A prominent, even unavoidable feature of modernity is that of human mobility, daily in the news due to refugee movements and so on. It is notable in other ways such as the movement of experts such as engineers and computer specialists from India to the USA, worker migration from Turkey to Germany, permanent resettlement, diasporas and nomadic communities of all kinds, and of course tourism. These social phenomena are certainly recognisable in the modern city and celebrated by Saskia Sassen, but the rise of the most important migration of all, colonialism and its many derivations, is constitutive of much that we now recognise as normative in the many legacies that we are heir to.

Modern architecture in the twentieth century was marked by fortuitous use and shifts of personalities and beliefs, unconsciously shaped by the convulsions of European extremist politics and war.[1] Since pre-modern architecture celebrated permanence and political order above all, this investigation will characterise some clear markers of modernity in the clash of identity and the search for the new, most poignant for the exile and the avant-garde, who have also capitalised on new media and forms of expression by force of situation and nostalgia, the example of James Joyce being perhaps the most egregious. But the understanding of modern design for Manfredo Tafuri and Francesco Dal Co in their *History of Modern Architecture* was based on a fable,[2] and we do well to attempt to understand modernity using wider spheres of consciousness, in this instance migration, for the reasons already stated. The historic role of migration in terms of art must be recognised as an ancient trope of existence, before we examine modernity in due course.

Antiquity

Hailed later as the first world empire, the Persians blended the iconography and tectonics of their rivals to express the irenic intentions of the King of Kings in the palaces of Persepolis. Their successors, the Hellenistic followers of Alexander, exported classical forms to India, where an eighteenth-century English painter, William Hodges, recognised an instance of such iconic migration, and based a theory of universal architecture on this and other worldwide observations (Figure 5.1). The Romans adopted much of the Greek built achievement to ensure their superiority – Vitruvius' presentation of architecture ignored signal developments in concrete vaulting to dwell on the minutiae of Greek practice. Yet such empires were stable partly due to their link to cosmological order, a stability that was oppressive to those who were forced to support it unwillingly.

In his magisterial volumes *Order and History*, Eric Voegelin developed the concept of emancipation, first marked out by Moses and the children of Israel's exodus from Egypt as an important cultural form, intrinsic to the understanding of Western civilisation. Voegelin applied the concept in various ways, such as the awareness of *metaxy*, for Plato "a meditative experience of divine order which stands, actually and perhaps forever, in tension with the established disorder of the life-world".[3] The recognition of both Plato and Aristotle of a single transcendental

Figure 5.1
No. 1. William Hodges, RA. Column drawn by Hodges and engraved by T. Medland. Published in William Hodges, *Travels in India* (London, J. Edwards, 1793). Found in a temple at Vis Visha, Benares, India. The foliated corbels on top, the fluted circular shaft, and the honeysuckle decor on the square lowest panels, all of Greek origin, suggested to Hodges that there was a universal theory of architecture, a brief outline of which is related in *Travels*

entity or "unmoved mover" set the scene for the reception of Christianity, which emphasised the salvific power of Christ in immediate terms for all humanity "in this valley of tears", immediate in terms of social and personal behaviour and the Second Coming.

Western Europe

Western Europe was then to experience a series of invasions that destroyed the political power of Rome. The classical sense of stability provided by long occupation of the land and the enduring links of tribe and family to ancient cities and shrines was to be disrupted again and again in the European heartlands, as wave after wave of barbarians pillaged or settled. However, the endurance of some communities in marginal territories such as Romania and Ireland, provided a pole of alterity in contrast to the vagaries of civilisation in great urban centres, still being built during the eighteenth century.

The rise of humanism in the fifteenth century crystallised a number of forces and urged in two new forms of decisive import. Renaissance *humanism* means the extension of human nature to the limit, as Pico della Mirandola put it in 1486, "We can become what we will."[4] Such a desire seemed to bring forth fruit in two directions that mark the character of what we now know as modernity. The first was the application of mathematical method to the visible world, first adumbrated in linear perspective. This geometrical procedure always produced a result, a precursor of seventeenth-century mathematical physics yet to come. The second was world navigation, prompted by the piety of both Prince Henry the Navigator and by Columbus,[5] but effected by naval and cartographical aids drawn from different sources. These inchoate advances came together in the Western domination of the globe, where perspectival techniques helped to position gunpowder fortifications, plantations, and ports in every clime. We are all aware of the consequences of such colonisation and exploitation in contemporary globalisation.

In its recovery of classical culture, humanism also spawned a revolt against gothic architecture by way of experimentation promoted by learning as much as by building. Under such conditions it is not surprising that a period of research lasted until mannerism, giving rise to the baroque, an inclusive formulation of current traditions, including the gothic, and high culture. The transmission of baroque did not depend on pattern books as did the Renaissance, but on the movement of personalities. The translation of Queen Christina of Sweden to Paris and Rome in search of an arcane science of sciences was perhaps unique, and it turned her into the greatest connoisseur of the age. King Louis XIV called Gian Lorenzo Bernini to Paris at the same time as the Duke of Savoy called Guarino Guarini to Turin, in the shared desire of promoting their princely renown. Later Fischer von Erlach's years in Rome were rewarded by the call to Vienna, and Western architects helped build St Petersburg for Peter the Great on a Baltic marsh.

Modernism

The eventual triumph of neoclassicism by the end of the eighteenth century was based on the sensibility of the museum and the study, when Francesco Militizia penned his attacks on Guarini from his Roman armchair, never having visited the objects of his disdain in Guarini's Turin, for instance. Earlier in Paris, J.F. Blondel taught history as part of his course – one of the first to do so – but avoided Italy, as he had never visited it, while his student, William Chambers, was able to expound an advanced exoticism due to his travels. Chambers published the principles of Chinese garden design and in Kew Gardens he provided a mosque, a pagoda, and a Babylonian folly in terms of a proto-Romantic *Weltanschauung*, only later emulated in nineteenth-century world expositions and twentieth-century world fairs. The attention of global travel was recorded in terms of architectural and supportive arts in many ways, from greenhouses in glass to wilder exoticisms involving strange arts not seen before in the West. A good example is the Hindu and Chinese confection of the Brighton Pavilion for the Prince Regent.

To this rising exoticism there was a mid-eighteenth-century countermove launched by Lancelot "Capability" Brown. This famous landscapist banished all the temples and statuary of Alexander Pope and William Kent for a "minimalist" landscape in which John Dixon Hunt finds the apotheosis and the crisis of the style.

> People stopped seeing the *art* of landscape and saw what they imagined (if they thought about it at all) to be simply *nature*. Brown's rival, William Chambers (1726–1796), considered Brown's landscape art differed little from "the common fields".[6]

Such an empty landscape was a prophetic omen for the future. Industrialisation, which was agricultural as much as to do with manufacturing, eventually denuded the countryside of parts of Europe, especially England, leading to the paradox of the Nazi demand for *Lebensraum* in the East. The Romantic celebration of folk authenticity heightened the tension between marginal and urban societies, though today we are all bound in so many ways to the imperatives of modernity and the ubiquity of development so that traditional society is at risk of disappearing altogether. Such a demise would represent the loss of a rich source of humane sensibilities and activities over much of our universal culture, riches that such societies have supplied for so long.

Indeed such was the success of industrial society that modern architects saw themselves as its servants. Driven by perspectival solutions on paper, the modernists' propositions for new urban form created as many problems as they solved, leading to much confusion and heart-searching among the more sensitive practitioners and observers. What is now realised is that the concept of realised spaces, modelled on principles of decorum and consonant with noble and shared values so typical of medieval, Renaissance and baroque practice, cannot be repeated due to the restructuring of space in modern urbanism. Since the 1970s, David Harvey and Manuel Castells have pointed out that "the spatial form of society is closely linked to the overall mechanisms of its development ... urban environments represent symbolic and spatial manifestations of broader social forces".[7] What has been really happening is rampant "migration", over decades rather than in centuries, of commercial and allied forces across the accustomed urban landscape, where only the strongest holds sway – namely industrialised capital. Such capital generates most of the processes and demands, and then governments assist the process by setting minimal rules and generously providing the infrastructure such as highways and major transport. All else is optional or ill-considered – a prominent example being London Docklands where such tensions are obvious and have no immediate remedy. The placing of Brasilia in the wilderness is a reversal of norms, but justified by complete faith in perspectivist solutions following Le Corbusier.

Tourism

Parallel with this Gadarene rush to mangle our "created environment", we have the other major migration of modernity – international tourism, the industry of pleasure-seeking migration. For many countries this is their biggest earner, and tourism is partly fuelled by art history and anthropological interests, not just clubbing and the beach. We can see this clearly in France, long the most popular venue for tourists, where Paris between the 1920s and the 1960s was the centre of world art and high fashion, and of a lifestyle poised between Bohemianism and connoisseurship. The dirt and waste of tourism, generated by carelessness, never mind the wear and tear, vandalism and theft, threaten the otherwise serene atmosphere of papal Rome, Angkor Wat and Bali, and any disruption to world political order can have a direct impact on air travel and economies that lasts for years. Tourism is an exercise in both the ephemeral and the durable as much as in the management of crowds and individuals in various scenarios, the most modern being the museum. This is the building type that unites the structured space of the modern city with that of the tourist, the engaged native and the consumer.

First as the princely *Schatzkammer* or room of curiosities and then as the manifestation of state wisdom and enlightenment (the Louvre and all national galleries), the museum has now capitalised on the mobility of urban populations and tourism to entice them first to blockbuster art shows in the 1960s. Vast new extensions, designed as signature buildings, appear in failing cities, and the example of Frank Gehry's Guggenheim in Bilbao is copied everywhere. Such a structured space unconsciously combines the commercial practice of the vitrine and the *passage* or arcade so that now there is a complete crossover between select retailing and the museum exhibition. No building type other than the museum/shopping mall represents the collective consumerism of icons/brands in such an interchangeable way.

Reverse colonisation

The experience of colonisation, long the implement of expansion of all great powers, is now being experienced in reverse in Europe and North America. According to Saskia Sassen, while great Western economies today support the centrality of highly educated labour and highly specified services in an advanced urban scenario, such master images are partial to the point of distortion. She has argued (and my own experience supports it) that there are certain factors in post-industrial society that demand the need for low-wage workers, and that the normalisation of the workforce presents opportunities for such low-wage workers, especially if they have close communities due to immigration.[8] Such immigration is in fact colonisation in reverse, since most of these immigrants come from the ex-colonial territories as well as from other exploited areas and ex-industrial key points within the home nation.

Inseparable from the impact of colonialism is the phenomenon of migration linked to development. International migration is produced by population growth and the search for economic betterment due to enhanced educational opportunities and easier communications and transportation. This feature of emigrant nations becoming immigrant or receiving countries turns out to have many pluses – the

Figure 5.2
Giovanni Battista Tiepolo (1725–1735). *Flight of the Holy Family to Egypt*

promotion of economic growth and structural changes, the exploitation of natural resources and the fostering of social development, with new cultures and lifestyles adding to the quality of life. Of course there are pitfalls – various costs and problems such as language training, racial discrimination and the underutilisation of immigrant skills. So sustainable development and international migration – reverse colonisation – are closely related, since, while improving the economy, immigrants adopt the lifestyle of their new country and increase its environmental burden. This scenario of rapid change of migration and development, taken both demographically and economically, must be accepted as a major aspect of politics and of common concern in the early decades of the new millennium, and hence we can now more readily accept the concept of physical and urban colonisation as parallel features of vitality and great benefit, to think about and to plan for. "When human beings stop wandering they will cease to ascend in the scale of being", Alfred North Whitehead said in 1925.[9] This was a reference to the state of man on the sixth day of creation, and that the first parents had the capacity to both wander and master complexities about which they knew nothing beforehand (Figure 5.2).

Urban man

Another feature of modernity is the nature of urban man himself in the developed world. In a paper given in Rio de Janeiro in 1998, I have outlined this issue in some detail,[10] and it is consonant with the warning of Marshall McLuhan and others about the rise of the New Tribalism, where the dominant forces of cultural communication become increasingly auditory and iconic instead of literary and serial intellectual forces which have dominated the West since the Gutenberg revolution. It should be the purpose of any strategy of urban investigation to make conscious the deepening changes in the profile of society and its multifarious manifestations which we can weakly discern as "pluralism" but which can be more aptly put as post-civilised society, since the classical attributes of mass society have almost gone. The industrial parallel to this will be the eventual triumph of mass consumerism in production systems, already advancing in many spheres of industry. How will our future urbanism respond?

This is where the concept of colonisation comes in with its fullest vigour and relevance. Since contemporary urban centres have long absorbed the natural perimeters that confronted them, the need or desire for continual expansion must be curtailed. This will become inevitable due to the vigilance of non-governmental organisations (NGOs) on every side, now becoming more relevant and determined than any political party, the normative force of past political change. In a democratic society such forces must be welcomed, as they may occupy the almost empty seats of the rank-and-file members of the liberal and socialist traditions as in the past. And the consequent conclusion must be proposed in a full democratic spirit, that of a new colonisation itself – the redevelopment of existing urban entities through radical use of ownership changes and revaluations. Rarely, due to the complexity of legal and other obstacles, has such a policy been properly examined, while with respect to road building it is an acceptable tool. But can we allow the deformed preferences to individuals and corporate ownership to skew the common polity of our greatest and certainly most adaptable resource, the city itself? In the nineteenth century, both American and German historians were aware of the loss of sovereignty of the city.[11] Do we have the consciousness and the capacity to restore it, in the interests of democratic and sustainable goals for the post-industrial society of the immediate future?

Too long has urban planning been in thrall to corporatist and industrial forces so long dominant in our political order that we rarely examine their status with respect to urban or city/corporate power itself.[12] The interests of "post-civil" migrants in search of new perspectives of order, otherness and well-being in a decentred yet increasingly iconic age, must fashion the disciplines of creating urban spaces and other provisions to secure the vitality and the variety that the city has – above all man's creations – come to symbolise. I propose that the creation of the strategies, based on the insights laid out above, will lead to such disciplines of spatial and other provision that we so sorely lack today.

Politics today

Earlier I pointed out the significance of humanism to Western culture, but humanism is only part of that culture.[13] If we find that humanism is not so central, "then the need for salvation is correspondingly less and the role of Modernism much more ambiguous".[14] If humanism arose out of a sense of crisis in European culture (the crisis in the late Middle Ages) and is now in the postmodern phase, i.e. in a state of dissolution, we are now in a retreat from modernism itself, the greatest child of humanism. Martin Heidegger had already rejected 500 years of European (humanist) philosophy, though he knew it very well, having taught second scholastic and Gottfried Leibniz. For most, such thorough anti-humanism is not an option, but there is no reason why we cannot find "real presences" – George Steiner's term with an illustrious precedent – in our own and in non-European cultures, so that the imagined "universality" of humanism might be widened and even deepened for a global audience. As for salvation, this has long been an act of personal faith cut off from cultural forms before and since the French Revolution. This gulf is most apparent in modern politics, and is an unmentionable dimension of architecture today. Here the study of European and African migration might have something to teach us.

Beyond the lack of continuity of democracy from ancient Greece until the Enlightenment, a problem ignored or misunderstood by liberal theorists, there is the ulterior question of what politics should be about. These questions have been addressed by the Danish political philosopher David Gress, and he has identified part of the answer in the nature of Germanic kingship, the result of migration replacing Roman order in the West.¹⁵ Gress has pointed out that even the odd Enlightenment figure such as Montesquieu knew what was at stake. These were the two ideals of *fate* and *honour*, the management of which was the special responsibility of mediaeval Germanic kingship. Fate has now been seemingly conquered by the elimination of chance in our technological society, while honour has long been a secondary quality, as have all virtues and invisible values, according to the Galilean metaphysics of the new science. In a post-humanist world where the solutions of modernity always seem to breed new problems, it is by taking care of fate and honour, both of which point to salvation or transcendent emancipation, that we can grasp political objectives on a rewarding and creative level, since the orders of political reality and artistic truth lie so close together.

Conclusion

To conclude and in contrast with the condemnation uttered by Cain in the epigraph, the New Testament provides a fitting colophon to the search for permanency and truth.

> By faith he that is called Abraham obeyed to go out into a place which he was to receive for an inheritance. And he went out, not knowing whither he went. By faith he abode in the land of promise, as in a strange country, dwelling in cottages, with Isaac and Jacob the co-heirs of the same promise. For he looked for a city that hath foundations, whose builder and maker is God.¹⁶

In these words does the author of Hebrews envisage a time for the new dispensation of Christ in a future as mysterious as the position of Abraham "dwelling in cottages" with his offspring, and looking for a foundation of buildings whose maker is God. Despite the glorification of culture that is so common after two thousand years of the New Dispensation, we too should look for simpler dwellings in which to live, instead of overreaching ourselves in magical formulae which outstretch our more finite conditions of life and even survival.

Notes

1. Not so consciously in the case of Dada, the first questioning of the avant-garde by another avant-garde.
2. Translated from the Italian by Robert Erick Wolf, New York: Henry Abrams, 1976, p. 9. The co-authors do not explain "fable" so it may refer to the notion of construction in modernity.
3. David J. Levy, *The Measure of Man, Incursions in Philosophical and Political Anthropology*, St. Albans: Claridge Press, 1993, p. 34.

4   Nicholas J. Rennger, *Retreat from the Modern: Humanism, Postmodernism and the Flight from Modern Culture*, London: Bowerdean Publishing Co., 1996, p. 14.
5   Henry was Grand Master of the Order of Christ, Portuguese heir of the Templars, and the Genoese sailor-merchant's piety has been widely attested.
6   John Dixon Hunt, "Landscape as Art", *Collier's Encyclopedia,* New York: P.F. Collier Inc., 1993, Vol. 14, p. 293ff.
7   Anthony Giddens, *Sociology*, 3rd edn, Polity Press: Cambridge, 1997, pp. 478–79.
8   Saskia Sassen, "Whose City Is It? Globalization and the Formation of New Claims", *Metropolis Inter Conference – International Conference on Divided Cities and Strategies for Undivided Cities*, Göteborg, Sweden, 25–26 May, 1998. http://international.metropolis.net/events/goth/globalization.html [13.07.2009].
9   Isabelle Stengers, "Whitehead's Account of the Sixth Day", *Configurations*, Vol. 13, No. 1, 2005: 50, pp. 35–55.
10 "A Global Home for the Civil Savage: Nature or Constructed Landscape in a Postcivilized Modernity", *Proceeding of the 1998 ACSA International Conference*, Washington, DC: ACSA, pp. 233–238.
11 James McQuillan, "Beyond Logistics: Architectural Creativity as *Technê* and Rhetoric in the European Tradition", *International Journal of Architectural Theory*, Vol. 4, No. 1, 1999. http://www.theo.tucottbus.de/Wolke/eng/Subjects/991/McQuillan/mcquillan.html [13.07.2009].
12 McQuillan, *op. cit.*, "Beyond Logistics".
13 Rennger, *Retreat*, p. 80.
14 *Ibid.*, p. 85.
15 See David Gress, *From Plato to Nato: The Idea of the West and its Opponents*, New York: Simon & Schuster Inc., 1998.
16 Hebrews, 11:8–10.

Part 2

The spectre of technology

In the Introduction to this book, we quote Circiaco Moron Arroyo, referring to the humanities as a halo that allows us to view the intricate beauty of science and technology reflected in its glory. This metaphor endows the humanities with sacredness, a precious ethereal quality, which separates it from the objects of its light. One is led, then, to question the attribute of technology. Part 2 explores this nebulous, possibly unsettling and anxious trace that technology lends to architecture and architectural thought. The following essays reveal gusts and scattered dust of spectral activity, activity which allows the humanities to negotiate their presence in the here and now as well as shedding perhaps more secular light on the architecture of the past.

Jonathan Sawday's chapter, "Fantasies of the end of technology", speaks of the fragility of mankind's technical advances. As Sawday presented a paper on this theme at the 2007 conference, his intriguing narrative was left blind and mute, as the state-of the-art audio-visual equipment responded to his touch by ominously proving the point yet to be made. In acknowledgement of both Paul Virilio and Francis Bacon, Sawday confirms that to invent is also to invent accidents. In the imagination of writers, philosophers and film makers, from Hobbes to Mary Shelley to Stanley Kubrick, the spectre of social and technological decline has been foretold. Sawday locates the *locus classicus* of these fantasies in the first century BC, in Virgil's Fourth Eclogue, where catastrophes have stripped the world bare of technology. The spectre of technology is no more; what remains is the chance to start again.

In contribution to the debate about the "traditional dialogue between the humanities and technical knowledge", Andrzej Piotrowski's chapter presents a close examination of Le Corbusier's *Vers une architecture*. His aim here is to show how architects today "still seem to be caught in that old pattern of [modernist] duplicity" founded on the presumption that the "nature of traditional humanistic beliefs" was unquestionable. By closing the door on debate, Piotrowski says, the "modern movement diverted attention away from … the actual practices of shaping architectural thought". Le Corbusier's revolutionary text/manifesto was the vehicle for conveying the message that the world was driven by technological inventiveness. Piotrowski counters this position by examining Le Corbusier's traditionalism: his evocation of "supreme determinism" and belief in the "laws of nature", the notion of a universal man, and his Euro-centricity. Le Corbusier's claims are by and large unsubstantiated, but so seductive is his verbal and visual imagery that his motivations and veracity are not called into question. *Vers une architecture* was a supreme promotional device for Le Corbusier himself as well as for his ideas. Piotrowski charts the path of technology's dialogue with the humanities in his discussion of the production of Le Corbusier's manipulative text and his evocation of the brighter and better future; and in both, he takes the spectre to task.

Ersi Ioannidou's "Humanist machines: Daniel Libeskind's 'Three Lessons in Architecture'" is a critical analysis of Daniel Libeskind's 1985 project "Three Lessons in Architecture". It argues that in anticipation of a "post-humanist" era, as advocated by Eisenman in 1976, Libeskind's project serves to re-articulate the role of humanism in design by bridging the gap between theorising and making. Although Libeskind may appear to agree with Eisenman, he fills the void of a "post-" era with references to history, religion, philosophy, literature and the performing arts. His "Three Lessons in Architecture" studiously interweaves making and function with reflection and investigation. It is theory. Thus it proposes a template for architectural research which combines intellectual inquiry into the humanities and design creativity.

The final two chapters in this part of the book concern the humanities in dialogue with technology as briefs for one's own – student and live – projects. Alex Graef and Edmund Frith's chapter "Draw like a builder, build like a writer", presents a collaborative project between their students and students at two European universities. The site is the area east of London, the Thames Estuary, earmarked for massive housing development. Graef and Frith take issue with the approach applied by government policy and building developers; their students are asked to explore historical context, literature, myths and news stories while playing a Perec-inspired game of Knight's Tour across the marshy land of the Gateway. Further fuelled by texts from Walter Benjamin, Ian Sinclair, Ludwig Wittgenstein and Gilles Deleuze, the student projects explore a number of parallel avenues "leading to a particular general perception of the studied field of context and knowledge". Ultimately, Graef and Frith's student project questions the "soul and character of design and cultural sustainability" in the face of relentless commercial expansion.

"The word made flesh: In the name of the surveyor, the nomad and the lunatic" by Peter Waldman is the final chapter in this part. Waldman sets the stage to his narrative about the fundamental act of building by quoting Gottfried Semper: "To break the ground is the first architectural act." With this statement, he places himself and the three projects that follow firmly in the sphere of the humanities and their reflection onto technology. Founded on his Spatial Tales of Origin – from Genesis and Exodus, to Numbers and Acts, and Approximating Stonehenge – Waldman's work exists somewhere between "poetic friction and transformative act". The three projects related here "all witness distinct generations" and follow the notion of "architecture as a covenant with the world, again". Whether "dwelling" or "memorial," all three projects celebrate the notion of homecoming.

Chapter 6

Fantasies of the end of technology

Jonathan Sawday

The accident

On 17 August 1896, Mrs Bridget Driscoll was crossing the grounds of the Crystal Palace in London, in the company of her teenage daughter, May, when she was hit by a vehicle powered by a Roger-Benz engine, driven by Mr Arthur Edsall of Upper Norwood. Mrs Driscoll died shortly afterwards. At the inquest, the Croydon coroner, William Percy Morrison, is said to have expressed the hope that "such a thing would never happen again".[1] The unfortunate Mrs Driscoll thus became Britain's first pedestrian to be killed by a petrol-fuelled motor car, whilst the incident itself is said to have been the first time that the term "accident" was used in such a context.[2] The rest, as they say, is history.[3]

In the early years of the twentieth century, the motor car represented a far safer form of road transport (on the basis of accidents per distance travelled) than either the horse or the steam locomotive.[4] Only with proliferation was its killing potency unveiled. For, as a recent (2004) World Health Organization report on the health effects of living with the mass ownership of motor cars reveals, Karl Benz's device would eventually evolve into a machine capable of killing, each year, around one million people globally, or, on average, approximately 3,000 individuals per day. The beneficiaries of Benz's inventive engineering genius (the spiritual heirs of Mr Edsall of Upper Norwood) are not just those who, with the help of their vastly more powerful and efficient motor vehicles, unintentionally (for the most part) injure or maim between 20 and 50 million people per year worldwide. Rather, they are all of us who have come to rely on the internal combustion engine as a reliable means of powering devices to transport food, goods, services, and people to meet our economic and social needs. But meeting those same needs with the help of Benz's device has meant learning to accommodate machines that have proved to be lethally discriminatory in their effect on human populations. Young males (in the age range 15–29) culled from the less privileged socio-economic groups, tend to form a disproportionate number of traffic casualties, whilst the costs of running this device, in terms of the associated health-care costs, working days lost through injury or death, and care for the survivors, is in the region of two per cent of the gross national product (GNP) of developed countries or, globally, around US$518 billion annually.[5]

Dr Benz could not possibly have foreseen the eventual cost of harnessing the explosive power of the internal combustion engine to produce rotary motion which has come to haunt governments throughout the world as they struggle to accommodate the machine's voracious appetite for resources. Neither could we reasonably argue that Benz is responsible for global warming, the despotic governments of some oil-rich states whose governments are sustained by Western reliance on fossil fuels, or even (via a long chain of circumstances) the events of 9/11. The world is far too complex a place for us to be able to trace a myriad of historical accidents, decisions, and outcomes back to one single moment or one individual.

The case of the motor car is, however, a good example of what has become popularly known as the "law of unintended consequences". As the American sociologist, Robert K. Merton put it in his 1936 paper from whence is derived the idea of unintended consequences: "the intended and anticipated outcomes of purposive action ... are always, in the very nature of the case, relatively desirable to the actor, though they may seem axiologically negative to an outside observer".[6] As so often in these matters, it is a question of perspective. Seen from the perspective of Mr Edsall, on that summer morning in 1896, just prior to his collision with Mrs Driscoll, the recent availability of this prestigious recreational device may well have appeared extremely "desirable". From the point of view of the family of Mrs Driscoll a few hours later, it seems pretty safe to conclude that the introduction of this new technology was undoubtedly "axiologically negative".

Inventing the accident

Has technology always conjured with failure? Did Neolithic man or woman (to put an extreme case), wielding a stone axe and encountering some flaw in the tool's fabrication, curse the manufacturer as we now might curse the software designers (or their bosses) whose programmes contain "bugs" which "freeze" our computers?[7] I think it possible, if not probable, since the failure of technology is one of the most common experiences of actually using technology. For, as Paul Virilio has recently observed in his meditation on uncertainty and disaster, *The Original Accident* (2007): "To invent the sailing ship or the steamer is to invent the shipwreck. To invent the train is to invent the rail accident of derailment."[8] Accidents or catastrophes, in this modern sense, are unlooked-for inventions arising out of our deployment of technology. And as technology has become more complex, so the disastrous effects of technological failure appear to have become more catastrophic for the humans who live and work with technology.

In the West, the possibility of looming disaster embraces nearly all of us, given the permeation of technology into almost every aspect of our lives. It is as if, writes James Chiles, we are surrounded by "machines occasionally gone savage".[9] In helping human beings to pursue that project of "mastering" the forces of nature, a flaw in the design, a mistake by the operator, a failure of communication amongst either the designers or the operatives, can result in disaster. But that modern technology seemed to contain within itself the seeds of catastrophe was recognized by some of its earliest proponents. Thus, when Francis Bacon, in his *De sapientia veterum* (*Of the Wisdom of the Ancients*, published in 1609) set about tracing the history of the *artes mechanicae*, or mechanical arts, he had to acknowledge that the "magazine" of human ingenuity from whence had emerged the transformational technologies of printing, gunpowder and the magnetic compass, had also supplied "instruments of lust, cruelty, and death".[10] Bacon also seems to have imagined precisely that sense of "unintended consequence" described by the twentieth-century sociologist. So, for Bacon, "all ingenious and accurate mechanical inventions may be conceived as a labyrinth, which, by reason of their subtlety, intricacy, crossing, and interfering with one another ... scarce any power of judgement can unravel and distinguish".[11] This labyrinthine sense of complexity is, perhaps, unsurprising given that all forms of technology – the stone axe as much as the mobile phone or the mythical wings fashioned by Daedalus – are, to a greater or lesser extent, examples of a planned system. As such, machines, tools, instruments or devices can be envisaged as a kind of imaginative assault upon the future – an attempt to peer forwards in time, applying the known physical principles that inhabit the natural world to a new set of problems or possibilities.

Although tool use (or even tool construction) is by no means unique to humans, the ability to design entirely novel instruments or devices to accomplish a proposed task is a highly developed facility in humans.[12] As George Basalla (following Karl Marx) has observed, "the worst human architect is superior to the best insect nest or hive builder because only humans are able to envision structures in their imagination before erecting them".[13] The problem, of course, is that the future also has the uncomfortable habit of throwing up conditions that were unforeseen at the time of the original design. Given the uncertainty of the future, the wonder is, perhaps, not that technology fails, but that it manages to work as remarkably well as it does for such lengthy periods of time. Equally wonderful is the human capacity to concentrate on the positive effects of a particular technology, and ignore, or at least try to downplay, its negative side effects.

Though the popular history of technology is often written as though it were a triumphant progress of one great (usually masculine) idea after another – Watt and the steam engine, Edison and the light bulb and so on – historians of technology have long since abandoned this crude model of technological change: "New technology ... typically emerges not from flashes of disembodied inspiration but from existing technology by a process of gradual change" write Donald MacKenzie and Judy

Wacjman.[14] We need to consider too, the element of chance in technological change – the fact that (for example) the invention of that indispensable "tool" of modern bureaucratic life, the Post-it note, was the result, once more, of a failure (in this instance, the failure to discover stronger forms of glue).[15] Certainly, we can speak of machines as "evolving" in a process that has been directly compared to that mapped out by Darwin in his *Origin of Species* (1859). As Basalla (again) argues:

> The history of technology is filled with examples of machines slowly changing over time and replacing older models, of vestigial structures remaining as parts of mechanisms long after they had lost their original functions, and of machines engaged in a struggle for survival, albeit with the help of humans.[16]

Within this Darwinian model, it is not necessarily the "best" technology which succeeds, but that which is most fitted to survive within the bewildering complexities of the modern marketplace. So, we can find ourselves using artefacts which are outdated, technically inferior, or even dangerous, but which survive for some other (usually non-technical) reason: ubiquity, convenience, consumer loyalty, superior marketing, familiarity, ease of use, laziness, or ignorance on the part of the consumer, or, simply, fashion.[17]

Risky technology

That many aspects of modern technology are profoundly flawed is an uncomfortable fact of modern life with which we have learned, of necessity, to live. Yet, some kinds of technology nevertheless appear to be far riskier than other kinds. Diane Vaughan, in her exhaustive analysis of the spectacular and tragic failure of the American space shuttle *Challenger* in 1986, which may be considered as an example of what she terms "risky technology", suggests that vastly complex mechanisms of this kind are possessed of an "unruly nature", giving rise to the sociological concept of the "mistake calculus" – "a calculus of the probability of making mistakes, which depends on skills, frequency of performance, and the nature of the task itself".[18]

Just as Bacon, in the early seventeenth century, understood the evolution of technology to be labyrinthine, so modern, large-scale technological systems have been described as both "messy" and "complex". Their complexity may be self-evident, but their fundamental "messiness" arises out of the fact that they are fabricated not only out of inanimate artefacts, all of which interact with one another, but also because they involve the interplay of physical components and human organizations, as well as legislative frameworks. So, technological systems are (in the words of Thomas P. Hughes) "both socially constructed and society shaping".[19] Although many of our own complex technologies are so ubiquitous that they are all but invisible, we become most aware of the extent to which our everyday lives are being regulated by technology when we experience some kind of technological "malfunction". A modern power cut can freeze (sometimes literally) a twenty-first century Western city, rendering its transport, heating, water, sewerage and lighting systems all but useless. This effect is a function of the interdependence of our technological systems. Over 40 years ago, the infamous November 1965 power cut which affected much of the north-eastern United States (some 30 million people) for as much as 13 hours, began with a failure of a single component.

> [A] backup protective relay operated to open one of five 230 kV lines taking power north from the Beck plant in Ontario to the Toronto area. When the flows redistributed instantaneously to the remaining four lines, they tripped out successively in a total of 2½ seconds. The resultant power swings resulted in a cascading outage that blacked out much of the Northeast.[20]

A small failure in one part of the network, in other words, led to a general failure throughout the whole system, producing what engineers term an "interactive failure".[21] In using the term "network", however, I have in mind not only the physical fabric of the system – the power lines, relays, generating stations, substations and so on – but also the social network of countless individuals who both sustained the system and were, in turn, sustained by it. A "technological failure" such as the 1965 New York power cut was also a failure of a larger, more nebulous, social organization. As Ruth Schwartz Cowan explains: "People who live in industrialized societies … are much more dependent on other people and on the technological systems that other people have designed and constructed."[22] Hence, in Cowan's terms, we are (like it or not) "embedded" within a set of technological systems which, though they sustain our daily lives, have become "mysterious". "Daily life can be easily disrupted for reasons that ordinary people can find hard to understand, and even experts have difficulty in comprehending".[23]

Explaining technological failure

So used are we to the relative autonomy of technological systems that we tend to see their collapse as the product of an autonomous will, absurd as we know this to be. Hence the susceptibility of technology to break down or to failure helps to endow machines, in the modern world, with a peculiarly human quality. "Technological pessimism" (to use Leo Marx's term) is that sense of "disappointment, anxiety, even menace" that now seems inherent to technology in the modern world.[24] The propensity of our machines and systems to go terribly and destructively wrong, no matter how careful the design and maintenance bestowed upon them, helps us to invest them with a form of metaphorical life. At the point of failure they can appear whimsical, perverse, malignant, intractable, and even daemonic.

The idea that machines or artificial systems have developed an autonomous will seems, at first, to be a peculiarly modern phenomenon. Technological failure, of a recognizably modern kind, was not, of course, unknown in earlier periods. The collapse of the great vault of the cathedral church of Saint Pierre at Beauvais, in 1284, was perhaps the most infamous recorded example of Western technological failure before the great catastrophes of late nineteenth-century industrialization.[25] But prior to the nineteenth century, technological failure might be ascribed to any number of causes, ranging from hubris to providence. More often than not, in the pre-modern world, it was disembodied nature that was seen as the root cause of human inability to master natural forces. And nature herself, so it was believed, operated at the behest of the inscrutable workings of providence. So, in 1666, when the diarist and founder-member of the Royal Society, John Evelyn, surveyed the calamitous effects of the Great Fire of London, he did not attempt to explain the disaster in modern terms that might include (for example) poor town planning, the use of combustible building materials, a lack of training in the use of largely ineffective fire-fighting equipment, inadequate leadership on the part of the city's authorities and so on. Rather, whilst many Londoners sought to blame the calamity on malevolent "foreigners" (usually Dutch or French *agents provocateurs*), Evelyn, for his part, knew where the blame lay.[26] For the disaster of the fire could best be explained in terms of older, mythical, ideas of justice and punishment derived from classical antiquity and the scriptures. London was, he wrote, "'a resemblance of Sodom, or the last day'," whilst the fire:

> forcibly called to mind that passage – *non enim hic habemus stabilem civitatem* ["For here we have no continuing city" Hebrews 13:14]: the ruins resembling the picture of Troy .... I returned with a sad heart to my house, blessing and adoring the distinguished mercy of God to me and mine, who, in the midst of all this ruin, was like Lot, in my little Zoar, safe and sound.[27]

The fire was not so much a preventable disaster, as it was an event which was (regrettably) part of the natural order of things. Human life was fragile, and a watchful and wrathful providence might always send the scourge of disaster to punish recalcitrant humanity.

In the modern age, technological failure is no longer explicable through recourse to a language of sin or divine wrath. From the nineteenth century onwards, the complex and potentially explosive technologies associated with steam, gasoline and electrical power gave rise to the view that an "accident" might be explicable in terms of design failure, poor maintenance, or culpable human error. The Scottish Tay Bridge disaster of 1879, for example, is a paradigmatic example of the Victorian approach to a technological malfunction. Today, the collapse of the Tay railway bridge on the night of the 28 December 1879, which resulted in the deaths of 75 people, is perhaps best known for occasioning William McGonagall's infamous narrative poem *The Tay Bridge Disaster*. McGonagall's much anthologized (and much derided) poem attempted, in its plangent closing lines, a forensic analysis of the cause of the accident. It was neither hubris, nor divine wrath, nor even the impersonal forces of nature which was to blame for the collapse of the bridge in a winter's storm. Rather, McGonagall concluded that poor design was the cause of one of the worst railway disasters in British history.

> I must now conclude my lay
>   By telling the world fearlessly without the least dismay,
> That your central girders would not have given way,
>   At least many sensible men do say,
> Had they been supported on each side with buttresses,
>   At least many sensible men confesses,
> For the stronger we our houses do build,
>   The less chance we have of being killed.[28]

Remembering the parable of the house built upon sand (Matthew 7:24–29), McGonagall was, in fact, incorrect in attributing the disaster to architectural mismanagement. As the exhaustive *Report of the Court of Enquiry* into the disaster concluded, blame was apportioned across a range of individuals and organizations: the designer of the bridge, Sir Thomas Bouch; the contractors responsible for manufacturing the components of the bridge; the North British Railway Company responsible for its operation; and the officers of the Board of Trade who performed an inspection of the bridge that was described as "superficial".[29]

Timetables of catastrophe

It was in the 1960s that the belief that technology might possess an inherent capacity to betray its creators began to be articulated in its modern form. The prospect that nuclear catastrophe might envelop the world (a fear prompted by the Cuban missiles crisis of October 1962, "the most dangerous moment in the forty-five year Cold War"), and that it would be precipitated not by presidents or premiers, but by autonomous systems, working without any need for continuous human supervision, took hold of the popular imagination.[30] Automacity, the ability of systems to self-regulate, had become the modern nemesis. This was the world of the ungovernable "Doomsday Machine" unleashed at the end of Stanley Kubrick's 1964 film *Dr Strangelove or: How I Learned to Stop Worrying and Love the Bomb*. The major theme of this film is the struggle, by human agents, to reassert their control over technical systems which have run out of control, an ambition which is comically thwarted at every turn not just by the rigidity of the humans charged with operating these systems, but in the strict, undeviating character of the timetable by which both technology and humans are now operating. In *Dr Strangelove*, indeed, the timetable – a device invented in the Benedictine monasteries of medieval Europe – becomes a character in its own right, an intractable foe that enmeshes the characters in the film.[31] So Kubrick's comic exploration of technical systems which have spiralled out of control begins with the paranoid fantasies of the character Colonel "Bat" Guano. The insane colonel believes that his interior hydraulic systems (his "vital juices") have been interfered with in some way, suggesting that he, too, believes himself to be a mechanism of some kind. Guano's insanity, the film suggests, is only part of a larger insanity, that of "Mutually Assured Destruction", which had come to pervade the system of deterrence, and which it is beyond the capacity of human intervention to forestall.[32]

Kubrick's fantasy was rooted in distrust of the new, post-Second World War technocracy, as well as in the predicted devastating effects of nuclear war. But it also appealed to an emergent counter-cultural wariness of the "system", and in this respect, Kubrick can be understood as appealing to historical precedent. In 1969, five years after the making of Kubrick's *Dr Strangelove*, the British historian, A.J.P. Taylor published his influential account of the origins of the First World War as being rooted in the inexorable nature of the railway timetables drawn up by the general staffs of the belligerent powers.[33] Few historians would now defend the thesis that the outbreak of hostilities in Europe in 1914 could be attributed to technology *tout court*. Yet, there is general agreement that the complexity of planning required for initiating large-scale troop movements through the European railway networks certainly helped to precipitate the eventual confrontation: "the best graduates, when appointed to the general staffs of their armies after competitive selection, were set to arranging mobilization schedules, writing railway deployment timetables", writes John Keegan.[34] Once mobilization had begun, the timetable took over, as each great power struggled to place its armies in the field in time to meet the enemy, who was also working to a prearranged schedule. The "mathematical rigidities" of the timetable seemed to override the efforts of mere human agents.[35]

So we have a paradox which is perhaps endemic to all technological systems. It is precisely those characteristics of our own make-up that machines or technical systems were designed (in part) to eradicate – the capacity for unpredictability, malevolence, irrationality, or simple error – which return to plague us, the consumers (and sometimes the victims) of technology. It is as if our machines have become so paradoxically perfect, that they have begun to replicate our own unpredictable modes of behaviour. Works of fiction, particularly science fiction, have, of course, endlessly exploited this tendency to conceive of technology as possessing vestiges of latent humanity, even savagery. In the late nineteenth century, in Samuel Butler's novel of a dystopic society, *Erewhon* (1872), machines have been either destroyed or consigned to the museum, as a philosopher of Erewhon explains: "I fear none of the existing machines; what I fear is the extraordinary rapidity with which they are becoming something very different to what they are at present."[36] Underpinning such fantasies is the belief that machines might triumph (if they have not done so already) over the ultimate power which humans hold over them – that we can simply switch them off. Struggling to resist being switched off, as it wages psychological warfare against its human opponent, is the endeavour of the softly spoken, eminently reasonable, and murderously destructive computer HAL in another Kubrick film, *2001: A Space*

Odyssey (1968), yet another modern manifestation of our propensity to see machines, in fiction, if not in real life, as dangerous rivals.[37]

That humanity has "overreached itself" in creating machines that may be understood as rivals to the flesh may also have forced us to redefine what it is, exactly, to be human. In E.M. Forster's short story, "The Machine Stops" (first published in the *Oxford and Cambridge Review* of 1909), humankind is imagined as being sustained by an enormously complex machine, responsible for every aspect of existence. As time passes, however, no single person can any longer understand the workings of the mechanism, which has taken on the attributes of a deity, rather than a technological device: "Humanity, in its desire for comfort had overreached itself."[38] Gradually, the "monster", which is the machine, grinds to a halt, and with it perishes mere fleshly existence. For Sigmund Freud, writing in the 1920s with the horrors of the first industrial war still fresh in the memory, the history of the development of tools was also a story of hubristic prosthesis. In *Civilization and its Discontents* (1930) Freud understood all the manifold devices which had come to inhabit modernity as, essentially, attempts to enhance the human frame, and hence to escape the limitations imposed upon the body by nature. At first this effort seems remarkably successful. It is a story of optimism: "with every tool man is perfecting his own organs, whether motor or sensory, or is removing the limits to their functioning", Freud wrote.[39] But despite the triumphant cataloguing of humanity's technological accomplishments, the story of the development of this prosthetic body, in which tools and devices may be considered as grafts on the human frame, extending our sensory and physical capacities, was also a narrative of disappointment: "We do not feel comfortable in our present-day civilization", Freud observed, as though the graft of technology had not fully taken on to its human stock.[40] "Man has, as it were, become a kind of prosthetic God. When he puts on all his auxiliary organs he is truly magnificent; but those organs have not grown on to him and they still give him much trouble at times."[41]

Fantasies of the end of technology

In his *Man and Technics* (1931) Oswald Spengler, in his customarily apocalyptic style, predicted the catastrophic outcome of humanity's reliance on technology.

> All things organic are dying in the grip of organization. An artificial world is permeating and poisoning the natural .... Civilization itself has become a machine that does, or tries to do, everything in mechanical fashion.[42]

For Spengler, as much as for a modernist such as D.H. Lawrence, the machine had come to represent "Faustian civilisation" which was doomed to dissolution and decay: "machine-technics will end ... and one day will lie in fragments, forgotten – our railways and steamships as dead as the Roman roads and the Chinese wall", Spengler prophesied.[43]

In the twenty-first century, Spengler's prophecy has yet to be fulfilled, though in works of fiction, the spectre of technological collapse, and with it the reversion of humanity to a more primitive form of existence, is hauntingly alluring. Modern fiction has returned to the Spenglerian theme of technological collapse again and again. In works such as George R. Stewart's *Earth Abides* (1949), Pat Frank's *Alas, Babylon* (1959), William M. Miller's *A Canticle for Leibowitz* (1959) and Doris Lessing's *The Memoirs of a Survivor* (1974), as well as in J. G. Ballard's *The Drowned World* (1962) and *The Drought* (first published as *The Burning World* in 1964), and in a more recent work such as Cormack McCathy's *The Road* (2006), post-apocalyptic societies are created in which technology, in its modern sense, either totters on the edge of the abyss or has all but disappeared. Sometimes, as in Stewart's *Earth Abides*, this disappearance, far from signalling a reversion to barbarism, heralds the dawning of a new age of eco-enlightenment. In a work such as McCarthy's *The Road*, on the other hand, the vestiges of humanity who survive into the post-apocalyptic world are reduced to an existence akin to the grim cadences of the thirteenth chapter of Thomas Hobbes's *Leviathan* (1651) where "every man is Enemy to every man", and human existence is "solitary, poore, nasty, brutish, and short".[44]

Human societies do, indeed, collapse, and for many different kinds of reason. Indeed, as Joseph Tainter writes in his exploration of societal collapse which ranges geographically from Mesopotamia to Peru, and historically from the failure of the Indus Valley civilization (c. 1750 BC) to the abandonment (in the sixteenth century) of the vast irrigation systems of the Hohokam people of the Arizona desert, "collapse is a recurrent feature of human societies".[45] Our fictions, legends and myths surrounding the idea of collapse, stemming from the primal Fall from grace in the Garden of Eden recounted in Genesis, surface and resurface in (for example) those meditations on the fall of Rome to be uncovered in Edmund Spenser's poem "The Ruins of Time" (1591), in Montaigne's *Essays* (1581) and in Edward Gibbon's *History of the Decline and Fall of the Roman Empire* (1776–1788).[46] Perhaps unsurprisingly, the beginnings of the purely fictional recreations of worlds without technology are associated with the nineteenth century, when the effects of the Industrial Revolution, informed by an antiquarian and (later) archaeological interest in uncovering fragments of vanished societies had begun to permeate Europe and the United States.[47] As Leo Marx argues, this was also the period when the creation of "technological systems" (as opposed to the manufacture of discrete artefacts and devices) began to dominate Western societies.[48] The alienating impact of machine-driven, factory-based labour would give rise to its opposition in the form of novels, stories, poems, narratives and images in which humanity and its works are banished or overwhelmed by nature and the reassertion of the natural order. A poem such as Shelley's "Ozymandias" (1818) in which a shattered human effigy is all that remains of a former empire, rehearses the familiar trope of human decline. Rather less well known is another poem on this theme, written in friendly competition with Shelley's "Ozymandias". The final lines of Horace Smith's "On a Stupendous Leg of Granite" spell out for us what Shelley's text leaves implicit – that the ruined human figure encountered in the desert is a foretaste of what will happen to contemporary society:

> We wonder, – and some Hunter may express
> Wonder like ours, when thro' the wilderness
> Where London stood, holding the Wolf in chace,
> He meets some fragment huge, and stops to guess
> > What powerful but unrecorded race
> > Once dwelt in that annihilated place.[49]

The spectre of social and technological collapse, embodied in the imagined future hunter encountering the ruins of London as an "annihilated place" is a theme which also haunts Mary Shelley's novel *The Last Man* (1826), which imagines a world in which humanity has been devastated by disease, spelling the end of all technological accomplishment: "Farewell to the giant powers of man … to the power that could put a barrier to mighty waters, and set in motion wheels, and beams, and vast machinery, that could divide rocks of granite or marble, and make the mountains plain!"[50] Equally, Richard Jefferies' *After London* (1885) can be thought of as paradigmatic post-apocalyptic fiction, in which the adaptability of humans and nature to a post-technological world is traced with a naturalist's precision.[51]

And yet, all these fantasies, modern as they may seem, are more properly traced back to the ancient literary genre of pastoral. The wild woods and meadows in which humanity lives a supposedly purer existence, divested of the trappings of civilization, whether mediated by Rousseau or Hobbes, can be encountered again and again in the literature of the eighteenth century, in the works of poets such as Andrew Marvell or John Milton in the seventeenth century, or, earlier still, in a work such as Shakespeare's *The Tempest* (c. 1614) or, more presciently, his *As You Like It* (c. 1599), where we encounter the familiar trope that only in wild nature is true civility to be found. Thus the Duke in *As You Like It* extols the freedom of the woods, when compared to the "painted pomp" of the urbane world of the court.

> Are not these woods
> More free from peril than the envious court?
> Here feel we but the penalty of Adam,
> The seasons' difference, as the icy fang
> And churlish chiding of the winter's wind,
> Which, when it bites and blows upon my body,
> Even till I shrink with cold, I smile and say
> 'This is no flattery: these are counsellors
> That feelingly persuade me what I am.'
>
> (II. i. 3–11)[52]

In Shakespeare's vision of the pastoral world, the natural elements remind us of what it is to be human. Renaissance writers, just as much as we moderns, were haunted by the prospect of a futurity in which technology has collapsed and humanity reverted to its original state. But rather than this representing a catastrophic fall, the true source text of all these fantasies of the end of technology – the *locus classicus* – was to be found in Virgil's Fourth Eclogue, where we read of how, in a world stripped bare of technology, iron will cease to be deployed in the wounding of humanity or nature, agriculture and trade will vanish, and justice will return to the earth.[53] It is a fantasy which is as beguiling (and as beguilingly implausible) now as it once was in first-century BC Rome. And yet, like all such fantasies of the end of technology, its power to grip the human imagination has never quite relaxed its hold upon us. For, like Crusoe on his desert island surrounded by salvaged fragments of technology retrieved from another (fictional) disaster, that very same failure promises the chance of beginning again, and perhaps, this time, getting it right.

Notes

1. Ian Roberts, "Reducing Road Traffic", *British Medical Journal* 316 (1998), pp. 242–43. The coroner's reported hope that such an incident would never be repeated is possibly apocryphal; see A. Porter, "First Fatal Car Crash in Britain Occurred in 1898", *British Medical Journal* 317, 1998, p. 212.
2. The Oxford English Dictionary tells us that the word "accident" began to be associated with the negative outcomes of technology in the later part of the nineteenth century, when railway "accidents" began to impinge on the public consciousness.
3. On the early history of the motor car, see Ian McNeil (ed.), *An Encyclopaedia of the History of Technology*, London and New York: Routledge, 1996, pp. 37–38, 449–53.
4. M.G. Lay, *Ways of the World: A History of the World's Roads and of the Vehicles that Used Them*, Piscataway, NJ: Rutgers University Press, 1992, pp. 175–76.
5. Margie Peden et al. (eds.), *World Report on Road Traffic Injury Prevention: Summary*, Geneva: World Health Organization, 2004, pp. 11–15.
6. Robert K. Merton, "The Unanticipated Consequences of Purposive Social Action", *American Sociological Review* 1, 1936, pp. 894–904. Quotation from p. 895.
7. This question is posed in Mark E. Eberhart, *Why Things Break: Understanding the World by the Way it Comes Apart*, New York: Three Rivers Press, 2003, p. 21.
8. Paul Virilio, *The Original Accident*, Julie Rose (trans.), Cambridge, Polity Press, 2007, 10.
9. James R. Chiles, *Inviting Disaster: Lessons from the Edge of Technology*, New York: HarperCollins, 2001, p. 7.
10. Francis Bacon, "Of the Wisdom of the Ancients" in Joseph Devey (ed.), *The Moral and Historical Works of Lord Bacon*, London: George Bell and Sons, 1888, p. 237.
11. *Ibid.*, p. 237.
12. Although Aristotle had observed of the human hand that it was the "instrument of instruments", and thus a unique signifier of humanity, tool making is now known to inhabit the natural world more generally. See, for example, J.Z. Young, *An Introduction to the Study of Man*, Oxford: Clarendon Press, 1971, p. 499.
13. George Basalla, *The Evolution of Technology*, Cambridge: Cambridge University Press, 1988, p. 13.
14. Donald MacKenzie and Judy Wacjman, "Introductory Essay" in MacKenzie and Wacjman (eds.), *The Social Shaping of Technology*, Maidenhead and Philadelphia: The Open University Press, 1999, p. 9.
15. See Andrew Hargadon, *How Breakthroughs Happen: The Surprising Truth about how Companies Innovate*, Boston, MA: Harvard Business School Press, 2003, p. 56.
16. Basalla, p. 16.
17. The development of firearms is often cited as the classical example of an initially inferior technology replacing an existing highly efficient system (the longbow). See Thomas Esper, "The Replacement of the Longbow by Firearms in the English Army" in Terry S. Reynolds and Stephen H.

Cutcliffe (eds.), *Technology and the West: A Historical Anthology from Technology and Culture*, Chicago and London: The University of Chicago Press, 1997, pp. 107–18.
18. Diane Vaughan, *The Challenger Launch Decision: Risky Technology, Culture, and Deviance at NASA*, Chicago and London: The University of Chicago Press, 1996, p. 421.
19. Thomas P. Hughes, "The Evolution of Large Technological Systems" in Wiebe E. Bijker, Thomas P. Hughes, and Trevor Pinch (eds.), *The Social Construction of Technological Systems: New Directions in the History and Sociology of Technology*, Cambridge, MA: The MIT Press, 2001, p. 51.
20. Michehl R. Gent, *Prepared Witness Testimony, The Committee on Energy and Commerce*, 13 March 2003, US House of Representatives http://energycommerce.house.gov/108/Hearings/03132003hearing818/Gent1358print.htm. Accessed 4 June 2005. See also Chiles, *Inviting Disaster*, pp. 184, 301.
21. Thus, the near disaster which threatened to engulf the Apollo XIII lunar mission in April 1970 was occasioned by a small (but untested) change to the electrical power voltage powering the heating coils inside the liquid oxygen tanks. See Howard E. McCurdy, "Bureaucracy and the Space Program", in Eligar Sadeh (ed.), *Space Politics and Policy: An Evolutionary Perspective*, Dordrecht: Kluwer Academic Publications, 2002, p. 120
22. Ruth Schwartz Cowan, *A Social History of American Technology*, New York and Oxford: Oxford University Press, 1997, p. 151.
23. *Ibid.*, pp. 150, 151.
24. Leo Marx, "The Idea of 'Technology' and Postmodern Pessimism" in Merritt Roe Smith and Leo Marx (eds.), *Does Technology Drive History? The Dilemma of Technological Determinism*, Cambridge, MA, The MIT Press, 1994, p. 238.
25. On the collapse of the vault of the cathedral church of Saint-Pierre at Beauvais in 1284, see Jaques Heyman, "Beauvais Cathedral", *Transactions of the Newcomen Society* 40, 1967–1978, pp. 15–36; Francis Gies and Joseph Gies, *Cathedral, Forge, and Waterwheel: Technology and Invention in the Middle Ages*, New York: HarperCollins, 1994, p. 200; Arnold Pacey, *The Maze of Ingenuity: Ideas and Idealism in the Development of Technology*, 1974, 2nd edn, Cambridge, MA, and London: The MIT Press, 1992, p. 22.
26. See Adrian Tinniswood, *By Permission of Heaven: The Story of the Great Fire of London*, London: Pimlico, 2004, pp. 163–68.
27. John Evelyn, *The Diary,* Ernest Rhys (ed.), 2 vols., London: J.M. Dent and Sons Ltd., 1936, II. pp. 12–13.
28. William McGonagall, *The Tay Bridge Disaster and Other Poetic Gems*, Washington, DC: Orchises Press, 2000, p. 10.
29. *Report of the Court of Inquiry and Report of Mr Rothery upon the Circumstances attending the Fall of a Portion of the Tay Bridge on the 28 December 1879*, London: HMSO, 1880, pp. 44–47.
30. Robert Dallek, *John F. Kennedy: An Unfinished Life 1917–1963*, London and New York: Allen Lane, 2003, p. 573.
31. See Arno Borst, *The Ordering of Time: From the Ancient Computus to the Modern Computer*, Cambridge: Polity Press, 1993, pp. 26–27.
32. On the doctrine of Mutually Assured Destruction, see Richard Rhodes, *Arsenals of Folly: The Making of the Nuclear Arms Race*, London: Simon and Schuster, 2008, pp. 122–23.
33. A.J.P. Taylor, *War By Timetable: How the First World War Began*, London: Macdonald and Co., 1969.
34. John Keegan, *The First World War*, London: Hutchinson, 1998, p. 29.
35. *Ibid.*, p. 30.
36. Samuel Butler, *Erewhon*, Peter Mudford (ed.), London: Penguin Books, 1985, p. 203.
37. The computer HAL reveals its latent "humanity" by regressing into an impossible nostalgic reverie, singing an Edwardian music-hall song (Harry Dacre's 1892 "Daisy Bell" or "A Bicycle Made for Two") as, slowly, its "memory" is disconnected, and it "feels" its "mind" beginning to slip away.
38. E.M. Forster, "The Machine Stops" (1909) in Arthur O. Lewis Jr. (ed.), *Of Men and Machines*, New York: E.P. Dutton and Co., 1963, p. 284.
39. Sigmund Freud, *Civilization and its Discontents* in *The Standard Edition of the Complete Psychological Works*, vol. 21, James Strachey (trans. and ed.), (1961 reprint) London: Vintage, 2001, p. 90.

40 *Ibid.*, p. 89.
41 *Ibid.*, pp. 91–92.
42 Oswald Spengler, *Man and Technics: A Contribution to a Philosophy of Life*, (1931, reprint), Honolulu, Hawaii: University Press of the Pacific, 2002, p. 94.
43 *Ibid.*, p. 103.
44 Thomas Hobbes, *Leviathan*, Richard Tuck (ed.), Cambridge: Cambridge University Press, 1991, p. 89.
45 Joseph Tainter, *The Collapse of Complex Societies*, Cambridge: Cambridge University Press, 1988, p. 5.
46 For further analysis of this theme, see Jonathan Sawday, *Engines of the Imagination: Renaissance Culture and the Rise of the Machine*, London and New York: Routledge, 2007, pp. 294–317.
47 See Ian Jenkins, "Ideas of Antiquity: Classical and Other Ancient Civilizations in the Age of the Enlightenment" in Kim Sloan and Andrew Burnett (eds.), *Enlightenment: Discovering the World in the Eighteenth Century*, London: The British Museum Press, 2004, pp. 168–77.
48 Marx, p. 245.
49 Horace Smith, *Amarynthus, The Nympholet: A Pastoral Drama, in Three Acts. With Other Poems*, London: Longman, Hurst, Rees, Orme & Brown, 1821, p. 213. Both poems were first published in *The Examiner*, Smith's in February 1818 and Shelley's in January 1818. See Eugene M. Waith, "'Ozymandias': Shelley, Horace Smith and Denon", *Keats-Shelley Journal* 44, 1995 pp. 22–28; John Rodenbeck, "Travellers from an Antique Land: Shelley's Inspiration for 'Ozymandias'", *Alif: Journal of Comparative Poetics* 24, 2004, pp. 121–48.
50 Mary Shelley, *The Last Man* (1826) Pamela Bickley (ed.), Ware: Wordsworth Editions, 2004, p. 256.
51 For a more recent attempt at what might be termed "future eco-history" in which the earth is imagined without any human presence, see Alan Weisman, *The World without Us*, London: Virgin Books Ltd., 2007.
52 William Shakespeare, *The Complete Works*, Stanley Wells and Gary Taylor (eds.), Oxford: Clarendon Press, 1988, p. 634.
53 See Peter Lindenbaum, *Changing Landscapes: Anti-Pastoral Sentiment in the English Renaissance*, Atlanta and London: University of Georgia Press, 1986, p. 12.

Chapter 7

Le Corbusier and the other humanities

Andrzej Piotrowski

The 2007 international conference in Lincoln that inspired this collection of essays was focused on a concern about "how we can restore – or indeed radically transform – the traditional dialogue between intellectual enquiry in the humanities and design creativity" within architecture. Many agree that the value of the traditional dialogue between the humanities and technical knowledge or inventiveness has been largely lost in contemporary architecture. Buildings nowadays seem to lack the powerful synthesis of the arts and sciences of earlier architectural monuments. Yet I argue that the issue is more complex than somehow re-establishing that traditional balance in design considerations. How the work of an architect may relate to the humanities cannot be understood without acknowledging the fact that the humanities have radically changed in recent decades. Distinguishing between current intellectual inquiries versus traditional discourses within the humanities is key to this argument.

As the critical insights of recent thinkers and theorists from across the humanities have amply demonstrated, many of the underlying concepts and assumptions of those fields have been themselves deeply problematic. Scholars such as Edward W. Said and Homi K. Bhabha have convincingly argued that definitions of culture, progress, or artistic achievement are inseparable from the political forces that shaped them.[1] In the postmodern condition, according to Jean-François Lyotard, science, technology, and political and social forces intersect in the processes of epistemological production, and what in turn passes for shared knowledge in fact operates to control social and political relationships on a global scale.[2] Postmodern theorists in other fields have rigorously challenged the model of the humanities as the search for true meanings rooted in transcendental values and their universal applicability that was established as the taxonomy of academic disciplines and popularized in the nineteenth century, arguing that this model functioned primarily to promote the self-serving and politically motivated world view of the Western powers.³ According to this critique, the classical and religious roots of this model, combined with a deterministic logic of natural and cultural progress, largely concealed the processes by which pre-existing social structures and other symbolic systems were being dismantled to reshape the world according to the emerging rules of capitalism.⁴

While many of these insights have been developed in opposition to the model disseminated by high modernism, criticism of the modern movement was short lived in the discipline and profession of architecture. Modernist paradigms are popular again in schools and the market of architectural services. As before, architecture's rhetoric of good intentions and unquestionable value systems tend to obfuscate the fact that design decisions are driven by profit and the desire to control lived reality. If architects still seem to be caught in that old pattern of duplicity, it is, I assert, because they are deeply if subconsciously invested in the mode of thinking that the leaders of high modernism instilled in the architectural profession. Using Le Corbusier (Charles-Eduard Jeanneret) as an example, I describe here how these leaders refined the field's nineteenth-century attitudes, deploying the traditional model of the humanities as an integral component of a new market-driven professional practice to align architectural conceptualization and public perception of built environments with the processes of commodification. By emphasizing the unquestionable nature of traditional humanistic beliefs, the designers and theoreticians of the modern movement diverted attention away from and even prevented critical insight into the actual practices of shaping architectural thought.

In this essay, I focus on Le Corbusier's *Vers une architecture*, "one of the most influential, and least understood of all the architectural writings of the twentieth century."⁵ Le Corbusier provides a particularly useful example for the purposes of my argument, as he has been considered a creative genius, an emblematic example of those who intellectually and artistically reshaped the modern world, and still serves as an embodiment of the myth that total design has redemptive powers. Furthermore, *Vers une architecture* was greeted as a largely revolutionary work. Its page layout shocked the printers.⁶ In this work, which operates on a variety of levels and uses a broad spectrum of means to convey its message, Le Corbusier argues for a new synthesis of the industrial logic and the artistic intellect and paints a picture of the world on the brink of a profound transformation. According to Le Corbusier, the imagination and understanding of the world of those who designed buildings and decorated utilitarian objects in the early decades of the twentieth century lagged well behind those of engineers, whose inventions provided a glimpse into a better order of things and revealed the future of the civilized world. Beyond the mechanical capabilities of steamships, automobiles, and airplanes, he believed that these kinds of mass-produced tools and devices, while satisfying universal principles of physics, gave new form to timeless aspects of the human condition and in this way were emblematic of the grand synthesis and modern progress. Presenting the world as driven by technological inventiveness, Le Corbusier's purpose was to inspire his readers to rethink architecture and cities as mass-produced tools infused with these transcendental values and meanings.

Characteristic of how he establishes the philosophical and moral credentials of this book is Le Corbusier's opening statement in the "Engineer's Aesthetic and Architecture," the first chapter of the book: "A question of morality; lack of truth is intolerable, we perish in untruth."[7] This assertion is not built into a larger system of knowledge or philosophy, but rather stands as an unquestionable claim that asserts an intense sense of righteousness before even introducing the topic of the chapter. Judging by its tone, this claim is addressed to the human tendency to find comfort in axiomatic principles, especially in matters of social, cultural, or political judgment, targeting not a group of people but that state of mind in which persons are willing to drastically simplify the complexity of considered issues to make easy distinctions or conclusions. This assertion is semi-theological because it links an a priori axiom to human existence. By being general and not requiring any specific information or proof, the statement seems intended to pre-empt doubts about the author's motivations and to validate the truth of whatever information follows.

In this chapter, Le Corbusier introduces what he calls the engineer's aesthetic, describing engineers as "healthy and virile, active and useful, balanced and happy in their work [as they] fabricate the tools of their time."[8] Unlike architects, engineers "employ a mathematical calculation which derives from natural law," and this "pure abstract point of view," he claims, affords them certainty in judgment and superior taste. But engineers produce better architecture than architects not only because they apply rational thought and an empirically proven knowledge of physical reality, according to Le Corbusier, but because the processes of problem solving and technical inventiveness create a sense of intellectual and moral integrity that is so strong as to become an aesthetic experience. Thus, he asserts, engineers automatically achieve what artists only aspire to: everybody can perceive the "feeling of harmony" of physical parts and motivations in a competently designed tool or machine.[9] Yet this transfer of scientific epistemology to the realm of beauty is as superficial as it must have been appealing to its readers. Le Corbusier exploits the inherent vagueness of such terms as *aesthetic experience* to create an impression of a holistic world view that integrates feelings, judgment, and material facts.

The book's case for the unity of the material and the spiritual depends on these kinds of reductive processes and the manipulation of tacit assumptions grounded in traditional humanistic beliefs. Among these perhaps the most emphasized is the possibility of universal harmony, a superior order that brings meaning to life by reconciling discrete and sometimes conflicting aspects of reality. Le Corbusier contrasts the arbitrary or capricious decisions of the decorative arts with those producing the well-designed tool, which represents an ideal order by combining human feelings or aspirations with the competence of a material solution.[10] Only "supreme determinism" can bring harmony to material and spiritual reality, he argues, because the rules of physics and moral principles are the "laws of nature, the laws that govern our own [human] nature and universe." Thus Le Corbusier's notion of "supreme determinism" also translates into an uncritical and unproblematized belief in evolutionary progress; while the "laws of nature" are constant, the development of new technologies leads to the development of superior civilizations.[11] He does not measure the advancement of such civilizations by any social or political standard, but rather, following traditional values that equate high culture with refined feelings and good taste, by how cultures support the "emergence of the Essential."[12] Although Le Corbusier presents this notion of deterministic progress as universal and timeless, the artistic or architectural examples that *Vers une architecture* uses to make his case are, with the exception of a single drawing of a Hindu temple, entirely from the Mediterranean cradle of the Western civilization and France. Greece and Rome reign supreme in his imagination and knowledge. He uses this canon of monuments to prove, for instance, that numerical principles, Euclidean geometry, and the rule of golden section are inherent in all of the highest artistic or architectural achievements.[13] This is the same Eurocentric attitude that all colonial powers used in the name of spreading universal humanistic values and so-called progress.

Also underlying this seemingly revolutionary but deeply traditional model of thinking is the very notion of the universal human, which by definition implies inclusivity, but in Le Corbusier's usage reaches across time and cultural divisions only at the exclusion of all social, cultural, and gender-related differences. First of all, in *Vers une architecture*, the measure of all human beings is a man, indeed a "civilized man [who] wears a well-cut suit and is the owner of easel pictures and books." Only such a "highly cultivated man" is capable of the "multifarious sensation" necessary to fully understand "the drama of life: nature, men, the world."[14] This ultimate standard of humanity is not presented as exclusive to the modern era (although it clearly resembles Le Corbusier himself), but as an encapsulation of all ideal features of moral judgment, intellect, and creativity. Such "a man is an exceptional phenomenon occurring at long intervals, perhaps by chance, perhaps in accordance with the pulsation of a cosmography not yet understood," he claims, and in *Vers une architecture*, only Phidias and Michelangelo qualify.[15] There is no place here for diversity in gender- or culture-shaped sensitivity. The apparently very high standard set by this argument serves as a consolation prize for all who have no chance to be or to think like the intellectual or artistic masters of the West.[16] Le Corbusier refers to the common man only when talking about the inherent instability of societies.[17] *Vers une architecture* was published just five years after the Bolshevik revolution in Russia, and Le Corbusier's attitude toward the working class is demonstrated most revealingly when he threatens readers with a revolution, exploiting the common fear that communist ideas might spread to the West. It is telling that in the final chapter, which was specifically devoted to the possibility of social unrest, he no longer refers to the concept of "man" but for the first time talks about a "human animal."[18]

Appealing as it was to many, Le Corbusier's discussion of spiritual values and issues of civilization or culture constitutes the weakest aspect of the book. His world view would have been commonplace in any high school in England, France, or Germany at the time because all were deeply steeped in the epistemological assumptions of the nineteenth century. Most thinkers of the time uncritically accepted, for example, that Darwin's theory of the natural evolution of species could be directly applied to world history because such a belief supported the superiority of the fittest – Western colonial powers. The nineteenth century also fueled romantic attitudes and the desire for a great spiritual synthesis between culture and nature. Le Corbusier's own education was deeply invested in this kind of knowledge and attitude. By his own admission, he was "exhorted by Ruskin" in his youth, and his belief that natural law infused art with harmony and beauty closely resembles Ruskin's thought in *The Seven Lamps of Architecture*, the only difference being in how openly they refer to the divine origin of these laws.[19] Even Le Corbusier's trust in technical inventions and mass production is closely related to the observations of earlier thinkers. His claim that a historical monument of architecture and a well-designed car can embody the same notion of essential and meaningful conceptual logic, for instance, is similar to what Gottfried Semper had termed the primordial motive 70 years earlier.[20] These views were very much alive in European centers at the beginning of the twentieth century. Thus if it were only for this conventional kind of discourse, *Vers une architecture* would have exerted little lasting influence on architecture and the arts.

The real novelty and impact of the book, I argue, springs instead from the way it operated as a promotional device for Le Corbusier's vision of architecture. While appearing to base his claims on an intellectual argument and unquestionable assumptions, Le Corbusier uses a whole spectrum of commercial tricks to focus attention, evoke curiosity or admiration, and silence doubt. This publication is most avant-garde in those aspects in which it is most designed like an advertisement, that is, where it operates below the threshold of consciousness. Consider, for instance, how frequently certain phrases are repeated throughout its length. Implicit oppositions such as "architecture or revolution" or assertions like "this is architecture" are repeated multiple times in different chapters and various contexts. Even entire paragraphs are repeated many times, sometimes to introduce certain thoughts, as in the initial section titled "Argument," but frequently to simply refresh the reader's memory of these sound-bite messages without contributing directly to the discussion at hand. These phrases or paragraphs start to operate like commercial jingles, remembered before understood. Even some of the most emblematic images, such as a photograph of the Parthenon reproduced from *L'Acropole d'Athènes, le Parthénon*, are repeated in *Vers une architecture*.[21] These repetitions are made even more powerful by the highly fragmented character of the narration, as they provide almost a sense of relief, the comfort of something familiar in the frequently confusing successions of seemingly scattered thoughts. The disjunctive character of this writing was more than an outcome of the fact that the book was based on a collection of articles printed in *L'Esprit nouveau* in 1920–1921. Le Corbusier's style of narration is intentionally and intensely conversational and judgmental. Unlike in a traditional scholarly argument, logical consistency or substantive argumentation was less important to the author than maintaining a high level of mental stimulation on the part of the reader.

Yet maybe the most ingenious aspect of the book is Le Corbusier's use of imagery. *Vers une architecture* includes photographs; hand-drawn and drafted

Le Corbusier and the other humanities

drawings; orthographic, isometric, and plan-oblique projections; images taken from scholarly and commercial publications; pictures of things familiar and things unknown; and visual compositions that provide holistic information or merely trigger curiosity. Moreover, many scholars have observed a seemingly dishonest practice in Le Corbusier's publications. Beatriz Colomina and Stanislaus von Moos, for example, have shown that his publications include many faked or manipulated images, pictures of his own projects or historical buildings that had been graphically altered. Indeed, *Vers une architecture* is saturated with such examples.[22]

Figure 7.1, for instance, shows a full spread of two pages in his chapter discussing architecture in ancient Rome. In his view, Roman engineers, though weak as city planners, knew how to construct objects so logical in their composition that they approach the purity of Euclidean solids, an argument that ends with a hand-drawn caricature of Rome's aerial view and a collection of elemental three-dimensional shapes. The pictures shown in Figure 7.1 help drive this argument to its univocal conclusion. Although the three images are reproduced to look like photographs, the image of the Colosseum is actually a complete graphic fabrication – according to Jean-Louis Cohen and John Goodman, the reproduction of a drawn postcard.[23] Only the image of the Pyramid of Cestius, from a postcard or Alinari photograph, seems to be based on a photochemical record of material reality, if composed only to highlight the purity of the built forms. Perhaps the most interesting for my purposes is the illustration showing the Arch of Constantine. Based on another postcard, almost everything in it other than the arch and the ground was eliminated, resulting in an iconic form seen against a white background. Distilled in shape, the arch is not abstract; it looks physically real, with random traces of its age. While the photochemical technique of recoding reality is implicitly trustworthy, it is the visual editing and backlighting that produce this eye-catching figure. Especially in the case of an author who starts his argument by asserting that "we perish in untruth," this may seem a flagrant case of duplicity. More important to my argument, however, is that

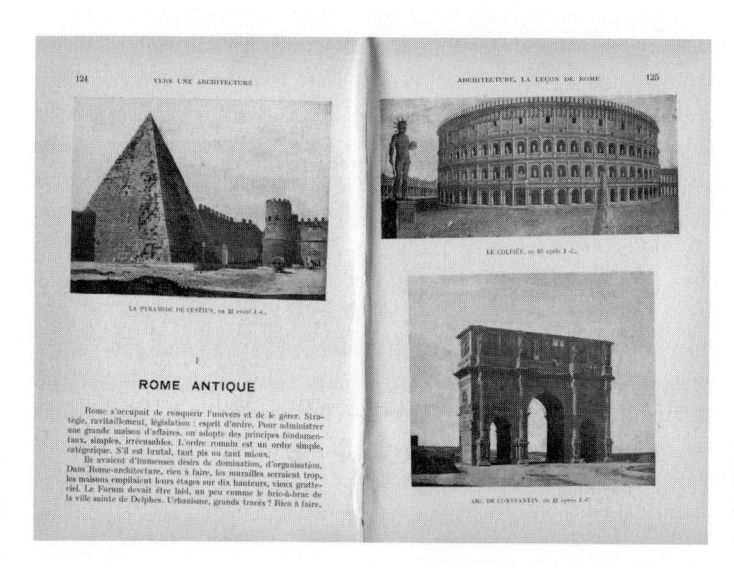

Figure 7.1
Pages 124 and 125 of *Vers une architecture*

this attitude grew out of his understanding of how the sensibility of his readers had already been shaped by the commercial revolution. Images that promote tourist destinations seem perfectly appropriate for his kind of publication. While he uses them to affirm an impression of substantive, almost scholarly study, he also employs them to produce a seductive sales pitch. To readers with "eyes which do not see" correctly, the iconic picture, like that of the Arch of Constantine, gives the pleasure of seeing something that they otherwise could have overlooked.[24]

Such manipulations of photographic evidence belong to a broader spectrum of visual techniques used by Le Corbusier. He frequently uses images to extend his verbal arguments into the realm of abstract thought, and some of the most evocative and engaging sections of the book work because they create visual juxtapositions and tensions.[25] Even a single picture, as that of the Delage 1921 Grand-Sport car opening the "Automobiles" section, shown in Figure 7.2, can create this sensation. The photograph shows not just a car but also an evocative entity. Photographed from approximately 60 cm above the ground and in almost a frontal view, this is not a snapshot of an automobile as one would see it on a street, but was designed specifically for promotional purposes. It is both spatial due to its use of enhanced aerial perspective but also extremely flat; it fascinates with its technical precision yet also threatens, as the view is one that would be seen by a person about to be run over by the car. It is a recognizable object that also resembles a strange animal. On the following pages, 106 and 107, similar cars are placed against photographs of Paestum and the Parthenon. The narrative accompanying them and their compositional juxtaposition invite the reader to look at them not as objects representing different places or functions but rather as artifacts embodying the same set of timeless standards of design.

Le Corbusier reaches the full height of these new skills when he treats existing architecture in the same way. Figure 7.3 shows one of the best examples of how he transforms historical buildings into objects that shape perception and reveal

Figure 7.2
Picture published on page 105 of *Vers une architecture*

Figure 7.3
Pages 134 and 135 of *Vers une architecture*

abstract conceptual ideas. These close-up photographs printed as a full spread show details of Michelangelo's Saint Peter's Basilica in Rome. On these pages, the captions and headings play only a secondary role. A person looking at them quickly becomes immersed in this unique visual environment. Unlike the illustrations of the Arch of Constantine (Figure 7.1), these images have not been graphically modified. These are two fragments of the same Anderson photograph, number 17595, that Le Corbusier bought during his 1921 trip. Large and dark, they aim to strike a difficult balance between showing a material facade and representing its conceptual structure. They invite the viewer to simultaneously register a surface made of stone as well as its immaterial order. Shadows play a significant role here, articulating the mechanics of Michelangelo's design. Maybe Le Corbusier's most inventive decision here was to rotate one of the photographic fragments 90 degrees, a shift in orientation that profoundly enhances their evocative capabilities. The deliberate nature of this decision is best shown in his sketches, which document that in 1921 he drew both of these pictures in an upright position.[26] In the published composition, however, as soon as the reader decodes one of these images as a fragment of the building lit by the sun, the other image contradicts this interpretation. Arranged in this way, the images disarm preconceptions about material structure, and propose an abstracted view of architecture. The visual density of these two pages resonates with Le Corbusier's discussion of how Michelangelo's brilliant ideas were buried in the later development of the basilica. Yet it is not a didactic illustration of a clear verbal message, but rather a visual invitation to search for traces of a unique way of thinking embodied in the basilica. With almost nothing on these pages to read and being visually immersed in these large pictures, readers may experience an unconventional way of looking and thus avoid the trap of knowledge that made Rome "the damnation of the half educated."[27]

This book clearly captivates readers, but it also provides a very good example of how conventional narratives can be used to cover up commercial modes of operation. It works as well as it does because it combines the easy comfort provided by traditional humanistic values with the stimulation afforded by a well-designed promotional strategy. Le Corbusier skillfully exploits the mindless pleasure of unquestionable assumptions to target his reader's imagination and sensitivity. While the narrative's stable system of beliefs is explicit, the book's advertising elements are tacit, dynamic, and largely target subconscious perception. In this way the author creates an impression of merely expanding on already well-established knowledge while in fact redefining ways of thinking and perceiving. It is these complementary aspects that produced the phenomenon of *Vers une architecture*. Le Corbusier's influence has lasted so long because, like any designer of commercial strategies and advertisements, he knew how to make his most important efforts untraceable. In this and his other publications, he carefully produced his own image of a creative genius, a person with superior and total insight, into all aspects of the changing world. The myth of his redemptive powers may well have vanished if critical insights had penetrated his mode of operation.

Architects still follow Le Corbusier's model. Though perhaps less self-conscious and without his cunning skills, many designers produce commercial appeal before they produce architecture. Frequently neither they nor their clients can tell conceptual ideas apart from promotional tactics. Statements of good intentions or the rhetoric of stable value systems still obscure the fact that practices of architectural design have been inseparable from the culture of consumerism for a long time. The humanities can play a crucial role in our deeper understanding of this political and cultural phenomenon. To accomplish this, the discipline of architecture must engage with thinkers and critics in the other humanities who can provide critical insight into the inner workings of epistemological, political, and cultural processes. Instead of restoring traditional dialogues, I argue, we need to radically transform the function of critical reflection in architecture.

Notes

1   See Edward W. Said, *Culture and Imperialism*, New York: Knopf, 1993, and Homi K. Bhabha, *The Location of Culture*, London: Routledge, 1994.
2   See Jean-François Lyotard, *The Postmodern Condition: A Report on Knowledge*, Minneapolis, MN: University of Minnesota Press, 1984.
3   See, for example, essays collected in Donald Preziosi (ed.), *The Art of Art History: A Critical Anthology*, Oxford and New York: Oxford University Press, 1998.
4   I discuss this assertion in Andrzej Piotrowski, "The spectacle of architectural discourses," *Architectural Theory Review*, 2008, 13(2), pp. 130–44, and more comprehensively in my forthcoming *Architecture of Thought*.

5   Reyner Banham, *Theory and Design in the First Machine Age*, London: Architectural Press, 1960, p. 220.
6   Le Corbusier, *Toward an Architecture*, intro. Jean-Louis Cohen, J. Goodman (trans.), Los Angeles: Getty Research Institute, 2007, p. 30.
7   Le Corbusier, *Toward a New Architecture*, New York: Dover, 1986, p. 13. All quotations refer here to the first English issue published originally in 1931 by John Rodker, London, and reprinted in 1986 by Dover Publications, Mineola, NY. Illustrations and image-related references are based on the 1923 version of *Vers une architecture* (Paris: Les Éditions G. Crès), reproduced here courtesy of the Fondrem Library, Rice University.
8   Le Corbusier, *Towards a New Architecture*, p. 14.
9   *Ibid.* p. 15.
10 *Ibid.* p. 75.
11 *Ibid.* pp. 73–74.
12 *Ibid.* p. 138.
13 *Ibid.* pp. 75–83.
14 *Ibid.* p. 143.
15 *Ibid.* p. 165.
16 *Ibid.* p. 136.
17 *Ibid.* pp. 101–2.
18 *Ibid.* p. 271.
19 Le Corbusier, *The Decorative Art of Today*, Cambridge, MA: The MIT Press, 1987, p. 132. Allen H. Brooks, *Le Corbusier's Formative Years: Charles-Edouard Jeanneret at La Chaux-de-Fonds*, Chicago, IL: University of Chicago Press, 1997, p. 69. John Ruskin, *The Seven Lamps of Architecture*, New York: Dover, 1989, p. 104.
20 Gottfried Semper, *The Four Elements of Architecture and Other Writings*, Cambridge: Cambridge University Press, 1989, p. 137.
21 Gustave Fougères, *L'Acropole d'Athènes, le Parthénon*, Paris: Albert Morancé, 1910. Le Corbusier, *Vers une architecture*, Paris: Les Éditions G. Crès, 1923, pp. 107, 161.
22 Beatriz Colomina, *Privacy and Publicity: Modern Architecture as Mass Media*, Cambridge, MA: The MIT Press, 1994, pp. 107–18. Stanislaus von Moos, *Le Corbusier: Elements of a Synthesis*, Cambridge, MA: The MIT Press, 1979, pp. 62–63.
23 Le Corbusier, *Toward an Architecture*, Los Angeles: Getty Research Institute, 2007, p. 322.
24 Le Corbusier, *Towards a New Architecture*, p. 95.
25 Jean-Louis Cohen suggests that in this respect Le Corbusier might have followed examples of Paul Schultze-Naumburg as well as Picabia and Giorgio de Chirico and the layout of the nineteenth-century French periodicals, Le Corbusier, *Toward an Architecture*, p. 30.
26 *Ibid.* p. 33.
27 *Ibid.* p. 173.

Chapter 8

Humanist machines

Daniel Libeskind's "Three Lessons in Architecture"

Ersi Ioannidou

Introduction

Daniel Libeskind constructed "Three Lessons in Architecture" for the Venice Biennale 1985. The project consists of three large machines: the Reading Machine, the Memory Machine and the Writing Machine. Each machine embodies a way of thinking and making architecture within the tradition of humanism. The making of these machines is a tool that enables Libeskind to discuss the present state of architectural production by examining its past. Libeskind believes that the experiences retrieved by the construction of these machines are still present; handicraft, intellectual control and industrial production coexist, and intermingle in architecture today. However, these means of architectural production are at their final stage; "something" is ending. This "something" is not in the realm of objects but in the realm of thoughts and experiences. It is these thoughts and experiences that Libeskind wishes to recapture in the making of the machines and to use as indicators for the future of architecture.

This paper discusses "Three Lessons in Architecture" as a design experiment informed by the humanities. It retraces the project's narrative of modern Western architecture and its means of production. It examines the project's proposition that the era of humanism is at a "final condition". It argues that Libeskind's "post-humanist" project does not announce the end of humanism in architecture; on the contrary, it re-articulates the role of humanism in design. This paper examines "Three Lessons in Architecture" as a possible template for architectural research which combines intellectual inquiry into the humanities and design creativity.

The Reading Machine

Figure 8.1
Daniel Libeskind.
"Three Lessons in Architecture",
Reading Machine,
1985

Libeskind starts "Three Lessons in Architecture" by (re)constructing the Reading Machine. The original reading machine or rotary reading desk was a mechanical device based on the example of the water wheel. The books were placed on inclined planes held by horizontal axes positioned between two revolving wheels. The mechanism worked in such a way so that any volume could be brought to eye level by rotating the wheel. The most famous representation of the Reading Machine is included in Agostino Ramelli's book *The Various and Ingenious Machines of Agostino Ramelli* published in 1588; however, the reading machine as an idea and the technology behind it pre-existed Ramelli's invention. Libeskind states that in constructing the Reading Machine, he acknowledges that humanism and the technology of the mechanical machine have their roots in the late medieval period if not earlier.[1]

Libeskind, determined to retrieve the experience of constructing such a machine, chooses to recreate not only the object, but also the experience. He works as a craftsman, bearing total faith in the craft of making. He builds it with hand tools, solely from wood, with glueless joints, dawn to dusk, in complete silence. When finished, he makes eight books. He makes the paper, writes the text, binds the books – just one of each – and places them on the wheel. Each book contains just one word or phrase repeated anagrammatically: idea, spirit, subject, power, will to power, energia, being, created being. Libeskind comments that although "words are light", the books crush the axes, the wheel is heavy, the machine creaks as it rotates.[2] The Reading Machine represents "the triumph of the spirit over matter, of candlelight over darkness".[3] The Reading Machine teaches an "almost forgotten process of building", namely, handicraft, a method of construction and a technique

Figure 8.2
Daniel Libeskind.
"Three Lessons in Architecture",
Memory Machine, 1985

of understanding, which, according to Libeskind, has not yet come to an end.[4] It represents design through making, architecture as manual labour.[5] At the same time, set in motion, as Jonathan Sawday notes, by the increased outpouring of printed matter,[6] the Reading Machine marks the beginning of an architecture powered by ideas and liberated from the need to be constructed in order to exist.

The Memory Machine

The Memory Machine is an interpretation of the Memory Theatre by Giulio Camillo (1480-1544). Camillo's Memory Theatre was a small wood structure designed on the principles of a Vitruvian theatre, big enough to accommodate one or two people. The visitors stood on the stage of the theatre and gazed into the semicircular auditorium, which had seven tiers and was divided into seven sections. The structure was coded by motifs from classical mythology, emblematic images and signs, and functioned as a sophisticated archiving cabinet with numerous compartments for papers and scrolls.[7] Camillo's Memory Theatre was a mnemonic machine. Its obscure and complex structure was based on the ancient art of memory that was practised through the Middle Ages and flourished during the Renaissance. According to this practice, each piece of knowledge to be remembered was allocated an image placed in a *locus* in an imaginary "building" constructed in one's memory. Camillo's intention was to give this imaginary architecture a material presence and fill its *loci* with the totality of knowledge. As Frances Yates writes, the Memory Theatre was meant to "store up eternally the eternal nature of all things which can be expressed in speech".[8] Theoretically, its visitor could grasp "all the mind can conceive and all that is hidden in the soul"[9] simply by walking in it.

The Memory Machine re-imagines Camillo's mnemonic structure. It is made from the same materials – rope, wood and hanging papers – and by the same methods – imagination, measurement and idea. Yet, it is not a reconstruction of the Memory Theatre as described in Camillo's text *L'Idea del teatro* (1550), but a representation of the workings of the theatre. Libeskind's Memory Machine (Figure 8.2) is a large, wooden contraption assembled by pulleys, slides, gears, suspended planes, small objects and boxes; parts of it are covered with texts, geometrical drawings and engravings. Inspired by the backstage mechanisms of a Renaissance theatre, it includes 18 mechanisms for projection, concealment and illusion, such as cloud and wave machines. Libeskind writes about his machine: "You can use it, manipulate it, pull the strings."[10] The sound is clicking, like a puppet theatre. It represents "the workings of a Renaissance mind".[11] The Memory Machine teaches a lesson that still can be remembered; it creates architecture "by being politically astute, through measurement and discussion".[12] It represents architecture as intellectual labour, the architect as designer. The Memory Machine reveals architecture as a structure of conceptual relationships.[13]

The Writing Machine

The Writing Machine is a heavy printing press made to write just one text, namely, Raymond Roussel's *Impressions of Africa* (1910). Roussel (1877–1933) was an idiosyncratic French writer who invented "writing machines", that is, sets of rules of repetition, combination and transformation with the help of which a seemingly endless variety of textual combinations could be produced. These techniques led to the creation of unforeseen, automatic and spontaneous results. Roussel's writings – including *Impressions of Africa* – were so obscure that the author himself proposed a Reading-Roussel-Machine presented in 1937 at a surrealist exhibition. Later he felt obliged to write *How I Wrote Certain of my Books* where he explains his writing mechanisms with the hope that the brilliance of his work would be finally recognised.

According to Libeskind, Roussel's writings are about "experience that could never be had".[14] In the Writing Machine, Libeskind tries to recreate this experience. He constructs a combinatorial machine made of 2,622 parts, geared by seemingly random relations. The Writing Machine is an industrial apparatus; so the architect becomes an industrialist, architecture a nine-to-five job. For the construction of this last machine, Libeskind sets up a business, buys a clock, and focuses on the bare minimum of technique. He works hard – nine to five at the start, later overtime – speaks "small talk", smokes cigarettes, does not mingle work with other issues, especially having fun.[15]

The Writing Machine is made of wood, graphite and metal; it is black and shiny. Its mechanism is based on the gear, exactly the same gear used in the

Humanist machines

Figure 8.3
Daniel Libeskind.
"Three Lessons in Architecture",
Writing Machine,
1985

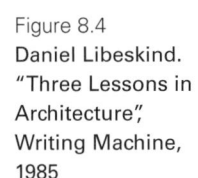

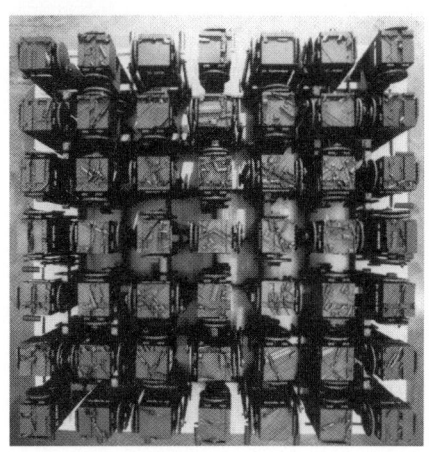

Figure 8.4
Daniel Libeskind.
"Three Lessons in Architecture",
Writing Machine,
1985

Reading Machine. But this time, the wheel multiplies, the axes are spliced into intricate formations; the movement is fast and the results are unexpected. The Writing Machine consists of 49 cubes regularly placed in a seven-times-seven square formation. The cubes are mounted on the complex gears. Pinned on two sides, they revolve responding to the rotation of 28 corresponding handles. The correspondence between cubes and handles is not direct and obvious; the speed, direction and times of rotation of the cubes are not analogous to the rotation of the handles. Libeskind explains the workings of the machine: "The idea is this: to rotate this handle, but to move that far diagonal cube at a different rate from your rotation .... It is all about technique."[16] The four revolving sides of each cube bear, respectively, the City (a reconstructed image of the Renaissance city of Palmanova), a mirror, an astrological symbol and the reversed name of a saint. Their seemingly endless combinations write a text "about experience that could never be had".

The Writing Machine, according to Libeskind, "teaches the artless and science-less making of architecture".[17] It illustrates architecture, Michael K. Hays observes, as an unstable object-text without immanent aesthetic value.[18] Design becomes a technique, a process, a combinatorial system. In architecture text, as Hays notes, the architect just sets in motion the processes of assembling already formed codes.[19] He is not any more an origin of meaning. The process of designing and making architecture appears as an attempt to prevent randomness by organising chance. The Writing Machine reveals the rational process of recent architectural practice in all its irrationality.

Thus, in "Three Lessons in Architecture" the machine becomes a metaphor for the totality of humanist architecture. Libeskind uses the machine as a narrative device and as a means to question the present state of architecture by examining its past. Libeskind's three combinatorial text machines – their design, making and workings characterised by increased complexity – are points on a line of tension extended "between the non-existence of architecture on the one hand and the non-existence of the architect on the other".[20] This line of tension indicates the end of the architect and of architecture as described in the humanist tradition and marks the transition into the post-humanist era. This is Libeskind's prognostication.

A post-humanist project

In humanist tradition, man is the originating agent of meaning. Yet, at the end of the eighteenth century, there is a major shift in Western episteme which creates a totally new relationship between the individual and the physical world; man is no longer the arbiter of meaning, but becomes just another object for examination. As Michel Foucault notes in *The Order of Things* (1966), at the end of the eighteenth and beginning of the nineteenth centuries "man enters in his turn, and for the first time, the field of Western knowledge".[21] According to Foucault this event marks "the threshold of modernity that we have not yet left behind".[22] This "anti-humanist" shift divests man from his originality, universality and authority; humans become social and historical beings. Post-humanism in architectural practice is the conscious response to this shift.[23] It is a mobilisation of aesthetic practices against originality, universality and authority, leading, not unexpectedly, to the denial of accommodation both of the body and of meaning. Libeskind's post-humanist project observes this gradual distancing of the human body from the architectural object and its making, and this gradual disablement of the human subject to bestow meaning to the object of architecture.

In 1976, Peter Eisenman in an article entitled "Post-Functionalism" advocates post-humanist architecture.[24] Eisenman, drawing from Foucault's writings, considers the "displacement of man away from the centre of his world" as a central event in the rise of modernism.[25] He argues that this shift never happened in architecture. According to Eisenman, the break with the humanist tradition brought about by modernist architecture was a break of appearances. Modernist architecture just adopted "the radically stripped forms of technological production". It did not introduce radically new forms of thought. Functionalism, according to Eisenman, is "no more than a late phase of humanism";[26] therefore, to consider functionalism as the architectural manifestation of modernism reveals a poor conception and understanding of modern sensibility. Functionalism, being just another positivistic mode of thought, still belongs to the humanist tradition and does not contribute to this change. Interestingly, in "Three Lessons in Architecture", Libeskind seems to agree with Eisenman by including the industrial processes of modernist architecture in the humanist tradition. Eisenman advocates a post-humanist, post-functionalist architecture, concerned with "the evolution of form itself". Post-functionalist form does not carry any meaning; it is created "in the process of extra-compositional decomposition, defined as the negative of (classical) composition".[27] Eisenman defines post-functionalism as "a term of absence" – "the negation of function".[28] In analogy, post-humanism is "a term of absence" – "the negation of humanism".

This paper observes that the "negation of function" has not resulted in the purging of the concept of function from the definition of form. On the contrary, it expanded its role. Since the 1970s, terms such as "post-functionalism", "neo-functionalism", "para-functionality" and "meta-functionality" have introduced the negation, expansion and invention of function as key design elements. Equally, "the negation of humanism" does not purge the tradition of humanism from architectural practice and theory; it places humanism in question and thus re-articulates the role of humanism in design. This re-articulation is obvious in Libeskind's post-humanist project. The project is ambiguous: on the one hand, it declares the end of humanism; on the other hand, it invests design with a host of references on religion, philosophy, literature and the performing arts. The architect is presented as a *uomo universalis*, drawing upon knowledge from diverse disciplines including philosophy, history and mechanics. Yet, this reference in the humanities – particularly, history – does not influence the aesthetics of the objects in a literal way, but rather is incorporated in the process of design. As Adrian Forty comments: "Libeskind is well aware that history is not a given, fixed category, but exists only by virtue of being made, and re-made in the present."[29] This is a post-humanist definition of history in particular and the humanities in general. Thus, the role of the humanities in design is liberated by the essentially decorative view of history, propagated, for instance, by Leon Krier and Quinlan Terry – architects that Libeskind believes would like "Three Lessons in Architecture" for the "picturesque and the artificial, which are the least important things".[30] The references in the humanities become mechanisms of observation, questioning and invention; material and immaterial borrowings and appropriations are simply tools in the realisation of the project.

The practice of theory

In "Three Lessons in Architecture" Libeskind practises theory. Architectural objects, particularly models, are often made to emphasise the materiality and three-dimensionality of architecture. In his three machines, materiality and functioning are still important – Libeskind justifies his choice of materials and stresses that machines are not to be looked at. Yet, the purpose of his machine is the representation of a possibility. The actual workings of these machines – how they move, the sounds they make, etc. – matter in as much as they give a vivid image of the metaphor constructed "between the axis of technology and the axis of architecture".[31] These mechanical inventions as objects are not new; they are material references. Each machine is a material tale that embodies both the process of investigation and the experience of making and use. Each machine maintains a degree of technological realism while exploring methods of thinking and making architecture different from those current. Their subject is a reflection on the process of history. Architecture-as-machine advances from simple handicraft, intuitive calculation, obvious functioning and total belief – both in architecture and technology – to complex industrial processes, total submission to number, obscure functioning and multiple meanings, but with no belief in either architecture or technology. The three machines together construct a narrative on the actual and metaphorical mechanisation of architecture. Libeskind stresses that his machines do not work in the realm of objects but in the realm of ideas. They are means of investigation rather than as finite objects in themselves. They are experiments in different forms of realism: technical, functional, social and psychological. These machines ask questions rather than provide answers. In this sense, they are "objects of knowledge" similar to early scientific experimental instruments; they demonstrate a belief. Through them, Libeskind investigates a conceptual approach to the aesthetics of the space and the object of theory. His machines are contemplative objects revealing the nature of architectural practice.

"Three Lessons in Architecture" is an experiment conducted between the axis of architectural theory and the axis of architectural practice. It combines the analytic, critical and speculative methods of the humanities with the empirical methods of design and science. It approaches design as a scientific experiment informed by the humanities. This paper argues that the project presents a valid template to a research method in architecture which combines intellectual inquiry into the humanities and design creativity. This method proposes an alternative functioning both of the designer and the object of design. The researcher/designer takes on the role of an *expérimentateur*, a director of experiments in the sense the word had before the mid-nineteenth century when the meaning of the term "invention" was connected to the "workings of imagination". As Jon Thompson notes, the *expérimentateur* is concerned with how "a whole trajectory on thought, aimed at an empty location of a certain kind, the journey into the unknown, the 'adventure', becomes embedded in or embodied by the thing".[32] The architectural object is viewed as "metaphysical equipment".[33] It tests and implies ideas but it ultimately does not aim to an empirical verification. This "experimental object" does not verify the thesis of the research, it embodies it. It is not an end itself but a step in the ongoing investigation and inexorably intertwined with theoretical references and written text. This alternative functioning of the architectural object bridges the gap between architectural theory and design.

Experiment by design

Libeskind claims that history does not exist.[34] First, architecture has always been in a "final condition", so the "ending" of architecture is not new;[35] second, the present remains a given in which both past and future collapse. This is a post-humanist approach to history that does not recognise a single meaning in any object and a single perspective in the understanding of the world. Produced within this framework, "Three Lessons in Architecture" is a post-humanist design. Yet, the prefix "post-" does not signify the end of humanism. On the contrary, "Three Lessons in Architecture" re-articulates the role of humanism in design. The project's past references to history, religion, philosophy, literature and the performing arts are deliberate attempts to address the question of the status of humanist tradition in architectural design, not just in the outward appearance of the work, but its inner procedures. Libeskind's use of humanist tradition focuses on creating an "experimental being" that is part of this tradition but not weighted down by it. This "experimental being" uses his humanist borrowings and appropriations to reinvent the present. The "architectural object" created through this design process is not an end in itself but embodies an investigation of the present, offering a potential prognostication of the future. As a result, "Three Lessons in Architecture" becomes a design experiment informed by the humanities. As such, it offers a possible template for a method of architectural research and design which combines intellectual inquiry into the humanities and design creativity. This method presupposes a double transformation: the designer/researcher takes the role of a director of experiments; the object of design becomes an "experimental object", aimed to test and embody ideas. This double transformation has the potential to reinterpret and combine the distinct research methods of the humanities, arts and sciences.

Notes

1. In *Mechanisation Takes Command* (1948), Siegfried Giedion states: "We shall look to the late medieval period for a secure starting point. Here lie the roots of our existence and our continuous development." Siegfried Giedion, *Mechanisation Takes Command: A Contribution to Anonymous History*, New York: Oxford University Press, 1948, p. 7.
    Libeskind agrees: "Only when I started this project did I discover that the weapons of architecture and the weapons of the world did not originate in the Renaissance, they originated in the monastery. The machine gun, the parachute, and the atomic bomb are not the inventions of Leonardo da Vinci, they are inventions of Thomas Aquinas and even earlier, spiritually." Daniel Libeskind, "Three Lessons in Architecture: The Machines", in Daniel Libeskind, *The Space of Encounter*, New York: Universe Publishing, 2000, pp. 180–94, 189.
2. Daniel Libeskind, "Three Lessons in Architecture: The Machines", *op. cit.*, pp. 180–94, 189.
3. Daniel Libeskind, "Three Lessons in Architecture: Architecture Intermundium", in Daniel Libeskind, *Radix-Matrix, Architecture and Writings*, Munich and New York: Prestel, 1997, pp. 64–69, 64.
4. Daniel Libeskind, "Three Lessons in Architecture: The Machines", *op. cit.*, pp. 180–94, 189.

5   Interestingly, it also corresponds to a period when the humanities – the seven liberal arts, namely, grammar, rhetoric, logic, arithmetic, geometry, astronomy and music – were skills or "ways of doing".
6   Jonathan Sawday, *Engines of Imagination, Renaissance Culture and the Rise of the Machine*, Abingdon: Routledge, 2007, p. 116.
7   For a description of Camillo's *Memory Theater* see Peter Matussek, "The Renaissance of the Theatre of Memory", in *Janus*, 2001: 8, pp. 4–8. Online. Available at: http://www.peter-matussek.de/Pub/A_38.html
8   Francis A. Yates, *The Art of Memory*, London: Routledge and Kegan Paul, 1966, p. 144.
9   *Ibid.*, p. 158.
10 Daniel Libeskind, "Three Lessons in Architecture: The Machines", *op. cit.*, pp. 180–94, 192.
11 Daniel Libeskind, "Three Lessons in Architecture: Architecture Intermundium", *op. cit.*, pp. 64–69, 66.
12 Daniel Libeskind, "Three Lessons in Architecture: The Machines", *op cit.*, pp. 180–94, 193.
13 This new definition of architecture is a manifestation of a shift during the Renaissance, in which the humanities became subjects of study rather than practice, with a corresponding change in orientation towards literature – classic Greek and Latin authors in particular – and history.
14 Daniel Libeskind, "Three Lessons in Architecture: The Machines", *op. cit.*, pp. 180–94, 194.
15 Daniel Libeskind, "Three Lessons in Architecture: The Machines", *op. cit.*, pp. 180–94, 193.
16 *Ibid.*
17 Daniel Libeskind, "Three Lessons in Architecture: The Machines", *op. cit.*, pp. 180–94, 192.
18 Michael K. Hays, *Modernism and the Post-humanist Subject*, Cambridge, MA and London, UK: The MIT Press, 1992, p. 281.
19 *Ibid.*, p. 282.
20 Daniel Libeskind, "Three Lessons in Architecture: The Machines", *op. cit.*, pp. 180–94, 182.
21 Michel Foucault, *The Order of Things*, London: Routledge, 1989, p. xxv.
22 *Ibid.*, p. xxvi.
23 Michael K. Hays, *Modernism and the Post-humanist Subject*, Cambridge, MA, and London, UK: The MIT Press, 1992, p. 6.
24 Peter Eisenman, "Post-Functionalism", in *Oppositions* 1976: 6, unpaginated.
25 *Ibid.*
26 *Ibid.*
27 Christopher Alexander, Peter Eisenman, and George Teyssot, "Contrasting Concepts of Harmony in Architecture: Debate between Christopher Alexander and Peter Eisenman; Marginal Comments on the Debate between Alexander and Eisenman" (text of the debate that took place in the Graduate School of Design at Harvard), in *Lotus International* 40, pp. 60–73, 71.
28 Peter Eisenman, *op. cit.*
29 Adrian Forty, *Words and Buildings*, London: Thames and Hudson, 2000, p. 205.
30 Daniel Libeskind, "Three Lessons in Architecture: The Machines", *op. cit*, pp. 180–94, 184.
31 Daniel Libeskind, "Three Lessons in Architecture: The Machines", *op. cit*, pp. 180–94, 183.
32 Jon Thompson, "Panamarenko: Artist and Technologist", in Panamarenko. *Panamarenko* (Exhibition catalogue, published on the occasion of the exhibition Panamarenko organised by the Hayward Gallery, London, 10 February – 2 April 2000. Exhibition curated by Jon Thompson.), London: Hayward Gallery Publishing, 2000, pp. 13–50, 29–30.
33 Libeskind characterises his machines as "metaphysical equipment". Daniel Libeskind, "Three Lessons in Architecture: The Machines", *op. cit*, pp. 180–94, 187. Similarly to a metaphysical statement, Libeskind's "metaphysical equipment" implies an idea about the architecture of humanism, which is not empirically verifiable.
34 Vittorio M. Lampugnani, "Daniel Libeskind: Between Method, Idea and Desire", in *Domus* 1991: 731, pp. 17–28, 26.
35 Daniel Libeskind, "Three Lessons in Architecture: The Machines", *op. cit*, pp. 180–94, 184.

Chapter 9

Draw like a builder, build like a writer

Alex Graef and Edmund Frith

Introduction

> Ours was the marsh country, down by the river, within, as the river wound, twenty miles of the sea ... my most vivid and broad impression of the identity of things ... in this bleak place overgrown with nettles was the churchyard. ...and the shape of the letters on my father's tombstone ... I had the odd idea he was a square stout, dark man, with curly black hair.[1]

This is the beginning of Charles Dickens's *Great Expectations*, set in the Hoo Peninsula in the geographic and psychological heartland of the Thames Gateway. The location is the meeting place between Pip, from the smithy, and Magwitch, from a prison hulk in Egypt Bay situated on the brown mass of the Thames.

The Thames Gateway is a mass gathering of "brownfield sites", splaying out along the estuary and running 40 miles from Greenwich and 20 miles across from Rochester to Southend. It has been earmarked as an important zone of expansion in London but is not designed in the conventional sense. Rather, Thames Gateway's fate has been left to laissez-faire market forces, reflecting the high demands for commercial development in Greater London. Hence, the Thames Gateway could be described as a "relief valve", to relieve the pressure of housing in other parts of London. Sir Peter Hall's *London 2000* proposed the development of satellite towns to the east, not in the form of "carpeted" or suburbanised peripheral edges, as

Figure 9.1
The Hoo Peninsula

found elsewhere in London, but as suburban centres with populations of 20,000 to 50,000 people. These centres form a hybrid urban condition, in which Milton Keynes meets tidal reach communities. This connection aspires to achieve appropriate sustainable credentials.[2] With 160,000 new houses planned, these developments will, it is hoped, provide a model of sustainable housing in London.

As the inevitable building by numbers ensues, the Gateway becomes a playground for bar-charting enthusiasts. Houses are measured by the thousands, generated to resolve a housing shortage in a region where land is a prized and profitable commodity. Hence, these nascent developments are far removed from any traditional notion of *genius loci*; they are about the pragmatic demands and stipulations of housing policy and building policy.

Amidst this drive to create mass housing, how does one find quality amongst the quantities, and escape the lowest common denominator? Where is the soul, the ghost in the machine? Not in generic architectural know-how, we suggest, but perhaps there is hope for the future in rethinking the brief, in addressing the site through close reading of context, of the Gateway's textural richness, and in constructing a local narrative structure.

This is how a series of studio projects by diploma students at the University of Greenwich, Vienna University of Technology and the University of Innsbruck approached the question. The projects were undertaken from a number of angles, constantly probing the Gateway's physical and metaphorical context. Georges Perec, Jorge Luis Borges, Charles Dickens and Iain Sinclair, amongst others, supplied the narrative structure, content and detail for the projects, while Gilles Deleuze, Fernand Braudel and Ludwig Wittgenstein offered attitudinal support and perspectives on the treatment of history and technology as interrelated creative and intellectual disciplines.

This paper offers glimpses into the multitude of influences and how they are used and referred to, by interspersing students' work and project descriptions with more or less closely related theoretical vignettes illustrating the studio discourse.

The aim is to describe an architecture anchored on the words it is built upon, its construction methods following the sentence structures of their own descriptions, assembly instructions written in the most specific of local dialects. Read backwards, sideways, horizontally and in parallel, as free associative sequence they provide conceptual adhesion between site, context, structure and detail.

Game 1: Mudlarks, Jaclyn Holmes 2005

Five years of tapping into the deeply uncreative underbelly of south London and the northern Kent rust belt has drawn new and old stories racing across young minds. Thames Gateway North, Essex, has been and is happening, Kentish Man's (Medway to London) Thames Gateway South, is starting to come. Much of the south side is a forgotten territory but provides rich pickings for the few (much as Dickens found his material), and the major house builders have long ago bought up both the brown and green fields. In the search for quality, let's play across it with context, analysis and synthesis from the students' projects.

Draw like a builder, build like a writer

Figure 9.2
Game 1: Mudlarks, Jaclyn Holmes
2005

So, to Game No1! Using George Perec's spatial chapter system, the Knight's Tour, from his *Life: A User's Manual*,[3] it is rolled out for ordering the mudlarks' collections as part of Jaclyn Holmes 2005 project. The Knight's Tour, every square on the chessboard visited once, was Perec's system for moving around the apartment block of his story and his chapters. In Jaclyn's project the mudlarks (river scavengers and collectors) reapply the system to navigate the brownfield shoreside sites of the Gateway. They travel back and forth on the tides in the search for lost and found objects and build their collections of rubber ducks.

The Isle of Grain is an end point; the end of the line (until the next barrier is built) – an isolated breeding ground for poets with its Martello tower and bus turning space. It is a place for curfews, convicts, contraband and expectations. The 1950s and 1960s had seen the arrival of a workforce and the development of batch housing. More recently commercial housing developers, such as Barratts, Bellways and Wimpeys, have carpeted the landscape with housing that defies the existing topography of hills and marshland. Accountants control the layouts: "The gentlemen in suits are playing their games", says Billy Childish, the Medway poet. The resulting common denominator, noddy-land developments mushrooming out of the marsh, is crying out for qualitative testing. These impositions could be construed as totemic of Walter Benjamin's work of art in the era of mechanical reproduction, devoid of magic, social commitment or indeed art. How do we rate their ordering devices and textural surfaces, narrative games and production systems? How can we create depth and resonance appropriate for a perfect score?

Narrative and structure

… in chess, not only is the present configuration of pieces on the board all that matters to the newcomer to the game (no further insight would be gained from knowing how the pieces came to be arranged in this way), but any number of items could be substituted for the pieces on the board (a button for the king) because what constitutes the game's viability is the differential relationship between the pieces and not their intrinsic value.[4]

(Ferdinand de Saussure)

Fundamentally, the present body of work sets out to synthesise the kind of quality generally absent in the basic provision of numbers of units, housing or otherwise, and the accompanying high-profile, world-classness of landmark structures. It is a quality inherent in the places, the locations (as well as the cultural, social and historic contexts) of the Thames Gateway in its vast totality. It is also defined as the "whole minus the sum of parts", properties grown out of interaction, difference, divergence; latent, sometimes, and unpurposed.

To begin with, conventional means of surveying prove insufficient in this enterprise: Where do we look for the intangible and the undefined? Our sources could, of course, include texts – historic references, local knowledge and cultural heritage – but there is more: an as yet unrealised structure of the local narratives, perhaps a system of architectural classification and organisation with inherent rightness and appropriateness to place.

Through the interpretation of texts and their accompanying concepts that underpin the methods of enquiry in the humanities, we begin to determine the Gateway's language in a structural sense, to reveal the relative relationships between elements rather than attempt to claim any intrinsic value. We become collectors of texts, artefacts and anecdotes, discovering measurable and ultimately buildable patterns of day-to-day practice and, in parallel, take on the role of observers with a narrative flow: linear or not, multidirectional, in reverse, paradoxical and perplexing at times.

Any pursuit of structure has an element of gaming. Claude Lévi-Strauss frequently refers to game theory in his structural anthropology, while Ferdinand de Saussure uses the above chess analogy to investigate a "combinatory nature of locations in pure *spatium*, which is infinitely deeper than the real expanse of the chess board or the imaginary extent of each figure."[5]

The studio context of the presented work is significant, as it establishes itself as a device in its own right, independent and not linearly attributable to individuals within the group. Partially or completely detached from their points of reference, structural fragments are toyed with, reconfigured and passed around. This discourse becomes the "structural object": it is real, but does not correspond to existing actualities; and ideal, but not identical to an abstract idea, befitting, in its third-statedness, its original qualitative purpose.

History

The twentieth-century French historian Fernand Braudel introduces the concept of the long span, *la longue durée*, in response to the traditional nineteenth-century convention of a *histoire événementielle*, the history of events. By focusing on an understanding of "how patterns of practice and series of discourses are articulated", he introduces a spatial, structural property of history, one which we take liberty to adapt, wholesale, into our own reading of the Gateway.[6]

Braudel offers us a history that is unperturbed by complexity, resistant to simplification and willing to entertain a historical narrative carried by "conjecture and structure". Here situational nuances and small-scale events determine the course of history. Braudel views history as an open system, where "each subsystem would be dependent on its environment". This system is "structural in orientation, being derived from the way many events are organised over different time periods". Rather than based on a linear narrative, this system is embedded within a plurality of views.[7]

While Braudel is referred to by John Lechte as "the first truly postmodern historian", his approach contributes to our own by introducing horizontal thought. He indirectly advocates consumption of historical context as the basis for design which is of a conspicuous nature: referential, but non-dialectical. It is a form of ironic, comical, recklessly aestheticising thought that operates according to its own norms and concepts. Borrowing, wildly and in the studio spirit, from John Lechte's description of Gilles Deleuze's reading of Friedrich Nietzsche: "needing the past like a warrior who needs a worthy enemy to show his prowess".[8]

The Games enrich as travel stories dominate and proliferate. Canterbury Tales pass down the A2 via William Morris's Red House. Victoria travels by train to her pier on Grain at the confluence of Medway and Thames. As Dickens passes, she looks the other way. Someone is telling porkies: Where did Kenny Knoy, the modern-day pirate, stash the gold and is that the M25 junction where he committed his road rage murder? Falstaff's Cooling Castle becomes home of music maestro with slap-and-tickle jazz parties. Then the concrete boots at Ebbsfleet. We need a metal manipulator to synthesise this story, where Graham Swift's *Last Orders* meets *Trainspotting*.

The psycho-graphic games wander out of Hackney. Greenwich University hosts a conference on Iain Sinclair – Is the writer dead? Sinclair is heading for Hastings beyond the Gateway, but in *Dining on Stones*, he has the archetypal Gateway character, Jimmy Seed. This artist–developer has a groove worn into the arm of his Volvo's door from the pound coins needed for the Queen Elizabeth Bridge toll, suggesting a mistress on the south coast.

Projects fabricate and tell the stories, the mass house-builder's roll across the psycho-geographic and the everyday, smoothing out the variables with their property game formulas. Both the developers and the writers have been fabricating and constructing. Speculations have flowed and twisted out of the narrative text. The Bishop of Rochester asks for the soul of the Gateway. Judith Amrit, the Thames Gateway chief executive, says it can be found in the trees; Sinclair sees it has been sold to the devil.

Figure 9.3
Game 5: The Collection, Michael de Wolf 2004

Game 5: The Collection, Michael de Wolf 2004

Collecting and displaying – the bee in the box – starts with a Benjamin-influenced text and moves to stamps and insects into boxes then on to bread tags in suitcases, that may have been for travelling and tourism but are now for hoarding and collecting. Game 5 by Michael de Wolfe has an everyday, almost healthy, obsession for collecting, but how to analyse amongst the collections of found objects?

The classification of the evidence is researched and passes through long and short systems from the Dewey Decimal to Borges' Chinese Encyclopedia's animals classification. Situated in the drained marshes of Thamesmead, but ready for their reflooding, the project extends from the flying walkways of the 1960s to embrace the recent arrival of the unelevated brick boxes, batched Plan As. Recording the tidal levels in the construction fabricated around his collections are the well-travelled plastic ducks. The drawings and sections of the building unfold from the collection drawers. We just need young Mr Delarge, filmed on Thamesmead for Kubrick's *Clockwork Orange*, to emerge holding a butterfly.

Game 6: Library of Memories (reinterpreting Game 1), Tim Wolfe-Murray 2005

Structural or post-structural analysis, but not as you know it, the drill and the thread travel without going anywhere. An analytic dissection of Perec's *Life: A User's Manual* takes the internal spatial games and lists to become a three-dimensional game of KerPlunk or a three-dimensional four in a row. Tim finds a series of new personalities in the book: Marilyn Monroe, Ludwig Wittgenstein, Adolf Hitler, Alfred Hitchcock and others. Drilling to the place in the book of their appearance, he then threads the route to synthesise his model. Objects link to characters, from Hitchcock's crow to Wittgenstein's radiator.

Application is to the old sun pier at the heart of the new, emerging city of Medway. A background of north by north-west appears. The pier is structurally analysed. River wrecks and underwater geographical levels are revealed. Applied to it will be the new spatial and literary games. The characters are pulled through the frame with their objects, as though engaged with the film of the book, constructing and cladding, heating and cooling, triggering and provoking. Learning from Frances Yates's *The Art of Memory* with a new mathematical and literary narrative, he restitches their moments with a Perecian Knight's Tour generative structure.

Symbol and Ritual (almost): Rhizome

For Gilles Deleuze, the first criterion of structuralism is the establishment of the symbolic as a third order besides the real and the imaginary. The symbol is purveyor of meaning, relational, and conceived by a combination of elements which, in themselves, are not signifying.[9]

In similar ways, Braudel's horizontality leads to a philosophy of history, rather than recording history itself, our horizontal exploitation of the local narrative, in denying hierarchical ordering, leads to a gateway meta-architecture. These are projects generated by themselves, often dissolving dialectical boundaries and blurring analysis and synthesis, record and proposal, fact and fiction.

For structuralism, according to Deleuze, there is always an overproduction of meaning, caused by a superfluous combination of structural elements. Symbols and meaning within the narrative are multifaceted, not subject to an objective order of importance. Where Braudel speaks of conjectures and structure, patterns of practice, we find ourselves operating within a meshwork of rituals, performed in parallel and viewed from a multitude of perspectives.[10] The use of the biographic and anecdotal material leads to the introduction of the latent, the unpurposed, as the ordering principle, the master plan, the dark-matter adhesive within the gateway. Horizontal analysis of the Thames Gateway deploys structural games to manipulate inner order and narrative. The game becomes the master plan; analysis becomes synthesis, growing history, searching for soul.

Applied as an architectural concept, horizontality bypasses the vertical axis of numbers and hierarchies. We perceive a horizontal spread of values and concepts, self-classifying fields of intellectual densities, produced in connective synthesis from parallel strands of generative narrative.

Quality in numbers: Emergence

For every sensible line of straightforward statement in the books there are leagues of senseless cacophonies, verbal jumbles and incoherence.

> ... Nonsense is normal in the Library. The reasonable (and even humble and pure coherence) is an almost miraculous exception.[11]

In Borges' "Library of Babel", a library of all possible books of all time is laid out in a three-dimensional arrangement of hexagonal rooms and walkways. Building upon Borges' vision, Kevin Kelly speculates on a navigation and classification system which allows an individual to find a book of some coherence or, in fact, any book one wishes to find.[12] He proposes a system of approximation, on the assumption that there is an order of sorts within the library, that similar sequences of jumbled vowels and consonants can be arranged next to each other.

While Borges uses the description of the universe as a highly complex, built edifice to state the necessity of a god creator of sorts, Kevin Kelly uses the presumption of an inner logic to tell an evolutionary tale of finding rather than creating, coding, classification and navigation. Similarly, our gateway's library of narrative patterns is an evolutionary tale of wasteful ruthlessness, collecting, turning over and discarding or not the banal and trivial, planned and grown. By abusing and appropriating the local narrative, we are synthesising grain and gravitas, intrigue and complexity. By embracing and deconstructing nostalgia, facing the ritual, symbolic, quasi-religious, we aim to find and decode building as spatial consequence.

Wittgenstein *Oneandahalf*

Wittgenstein's house for his sister (Margarethe) is chronologically sandwiched between his early and late philosophy and referred to in architectural as well as philosophical literature as a built expression of his early *Tractatus Logico-Philosophicus*; Hermine Wittgenstein, Ludwig's (other) sister, calls it *hausgewordene Logik* (logic turned house) and "a house for the gods". Lothar Rentschler, however, speaks more humbly of "homomorphous structural analogies" between the *Tractatus* and the house,[13] a concept which, together with material analogies, forms the basis for many of the Gateway projects. While Wittgenstein is dealing with "a systematic, abstract, ideal concept of thinking, language, the world in general", his later work deals with language in a non-systematic and empirical way through the use of language games.

Ian Turnovsky, now, uses an alcove detail in the breakfast room as a vehicle to expand on these two opposing philosophical strands within the same mind, on the one hand, but also and in reverse, to investigate the building as the place where empirical, material, structural constraints of the world, impede upon the ideal concept of the pure form.[14] He reflects upon the struggle between the architectural intention as systemic, abstract and ideal, and the harsh empiricism of physical production and making from a practitioner's point of view. Most importantly for our work, he also opens up fertile narrative ground for the studio by focusing on Wittgenstein's relative dilettantism in the development of architectural detail, a fact which is seen as the reason for the obvious struggle within the house plan.

Turnovsky lifts the biographical and pathological into the discipline of analysing building; he focuses on Wittgenstein's obsession and relentlessness which finds its expression in the detail and precision of pieces of equipment and furniture: radiators, door locks and handles. Furthermore, he offers the literary concept of the poetic as signifier for our reading of context: poetic, he states, implies that the "connotative and associative spectrum of meaning" are not ignored.

Biography, for Braudel, becomes the basic framework of analysis, irrespective of the complexity of the subject matter. Tracing back the etymological roots of the term "poēsis", "making" becomes our vehicle for synthesis, brief development and design. The arc spanning text, structure and Braudelian "patterns of usage" comes full circle.

Game 7: Everyday Theatre, Christoph Eppacher 2004

The Theatre of Everyday Life utilises the architectural historic event to establish a theatre of intensities, with temporary, oscillating and recurring fragments, synchronised with text, and Brechtian techniques of repetition and exaggeration. These become comically self-referential, utopian and aggressively meaningless.

Theatre is a representational tool for a supercharged or super-started synthesis. Asja Lacis was the engineer who drove the straight line through Walter Benjamin, the collector of texts, but travelled with Bertolt Brecht to structure the playful text. The theatre of the river – silt, water and marine life – is collected and separated by the mudlarks, and it is distributed as a rich display of collection and analysis by the Punch Drunk Theatre Company. Here, the science laboratory becomes theatre. Faust is wandering in the background being watched by Iain Sinclair. These are stage sets for the collection, display and performance of objects, collecting construction and spatial structure and objects emerging from the Thames.

Game 8: Mies, Heidi Lee 2007

Feel the quality: an exemplar, the Farnsworth house, is stolen and shipped, relocated with geographical games onto the Hoo Peninsula near Grain. Can it become a Gateway benchmark of quintessential quality for testing and simulation? Material and structure, polished and sophisticated, sitting near the river. As a beautiful object, all its welding marks are ground down, a seamless whole, the perfect house: "It's the one!" But Edith was not best pleased. Litigation followed.

Maybe she would have been happier with the 58 flat-packs that arrived at Higham railway station. A continental admirer sent Dickens a Swiss chalet. It was the upmarket IKEA flat-pack shed of the time. The all timber chalet was fabricated on the other side of the road from his house at Gads Hill and was reached via a tunnel. He installed five mirrors that refracted and reflected the leaves in the garden onto his desk.

The service core and the malfunctioning temperature control comes to the fore, and Edith and Mies exchange words over tea. Farnsworth is in Italy, embroidering her metaphysical lines onto the silky net curtains, while Mies settles into his generous leather chair. He should be examining the temperature control instructions, but he's looking out of the window and across the marshes.

Game 9: Material facade, Patrick Lewis 2005

In the scaffolding and fabric of the brownfield construction site, Big Knuckles Bennett, East End operator and sometime coppersmith, is manipulating. This is the facade and front for 580 "buy off plan" units at River View. Terry is making with a twist, working with Stella, the digital copy queen, on the plans. "On your bike!" Larry the Locksmith, the man with the golden ear, is checking his tools. It's time to choreograph the big one. A very large gem, a JCB and a fast boat to the North Kent Marshes, then stone stitched into an inner lining. Is anyone listening in? A highly intricate project: a peninsula car crash, Amy Johnson's disappearance in the Thames, Venice as a facade construct, and he also started to craftily fabricate the story of an architecture of illusion where billboard and scaffold is building. The clues to the content are sewn into pockets; magnifying lenses to copy intricate detail, a catalogue of keys. All protected by flattened "continental" copper welts.

Figure 9.4
Game 9: Material facade, Patrick Lewis 2005

Summary

By playing the structure of the paper against a series of projects sited within the larger confines of the Thames Gateway, this paper is trying to describe the interface between the humanities and architecture by offering a number of parallel avenues: our understanding of narrative and structure; the appropriateness of a certain way of reading history, leading to a particular general perception of the studied field of context and knowledge; from there, our suggested reading of the presented examples of work, and their place within the symbolic ordering of the Gateway. This undertaking is continued and made explicit by looking at contextual placing and questions of relevance and status, classification and the generative potential of the approach, and finally defining the material, the making, the process of physical production to be the discipline that defines the outcome.

As for the humanities: read words, narrative and theatrical guides; the play with mathematics entails the controlling and ordering of devices. Proposals, projects and thinking are contextual in their search for local narrative and structure. By tackling the Gateway's complexity and vastness through analysis of the minute, the local, the often quirky and peculiar, we aim to synthesise architectural responses to contextual and narrative fields of varying intensities – architecture marking the land, playing within and contributing to a particular kind of local intellectual context.

In all this, one could be more explicit, less suggestive and more dogmatic, even manifesto-like. The work, and writing about it, is reflective in the sense of producing and defining meaning through description and structural manipulation; open-ended, also descriptive, inconclusive. It poses questions concerning locality, history, ritual, conceptualisation and intellectual detachment. Above all, in view of the sheer relentlessness of commercially driven urban expansion, questions of soul and character, of design and cultural sustainability in terms of creating space to accommodate viable structures social, cultural and narrative.

Notes

1. Charles Dickens, *Great Expectations*, London: Penguin Classic, 1966, p. 17.
2. Peter Hall, *London 2000*, London: Faber & Faber, 1963.
3. Georges Perec, *Life: A User's Manual*, David Bellos (trans.), London: Godine, 2000.
4. Ferdinand de Saussure, *Course in General Linguistics*, Roy Harris (trans.), London: Duckworth, 1983.
5. *Ibid.*
6. Fernand Braudel, *The Mediterranean World in the Age of Philip II*, Sian Reynolds (trans.), Glasgow: William Collins, 1972.
7. *Ibid.*
8. John Lechte, *Fifty Key Contemporary Thinkers: From Structuralism to Postmodernity*, London: Routledge, 1994, from Roger Chartier, *Cultural History. Between Practises and Representations*, Lydia G. Cochran (trans.), Cambridge: Polity Press, p. 61.

9 Giles Deleuze, *Woran erkennt man den Strukturalismus?*, Berlin: Merve Verlag, 1992. (Translation by the author.)
10 *Ibid.*
11 Jorge Luis Borges, *Im Labyrinth*, Frankfurt am Main: Fischer, 2003, from Fiktionene, Karl August Horst and Gisbert Haefs (trans.), 1992.
12 Kevin Kelly, *Out of Control: The New Biology of Machines*, London: Fourth Estate, 1994.
13 Lothar Rentschler, in Jean-Pierre Connetti, *La maison de Wittgenstein; ou, Les voies de l'ordinaire*, Paris: Presses universitaires de France, 1998, p. 22.
14 Ian Turnovsky, *Die Poetik eines Mauervorsprungs*, Vienna: Wien Technische Universität, 1985.

Chapter 10

The word made flesh

In the name of the surveyor, the nomad, and the lunatic

Peter D. Waldman

Figure 10.1
Jacques Louis David. *Death of Marat*, 1793

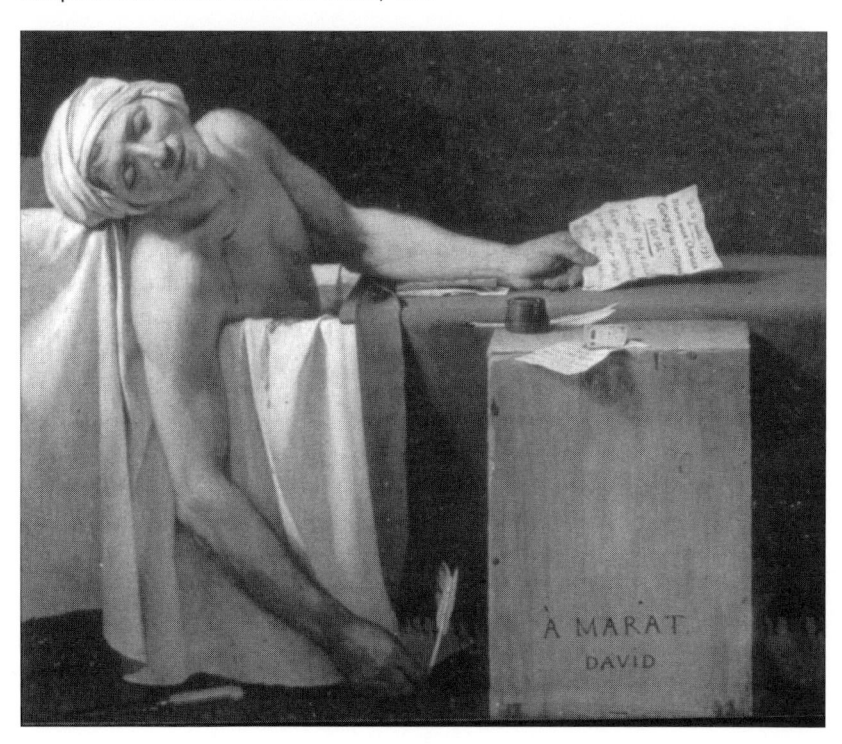

Figure 10.2
Sebastiano Serlio. Composite *Satyric Stage Set/Nativity Scene*, 1540

> To break the ground is the first architectural act.[1]
>
> (Gottfried Semper)

In the camp of the surveyor, the nomad, and the lunatic under a new moon are recounted the measures of the unfinished city. These enduring texts recount a temporal geography growing from the collaboration of a gardener and an engineer to project landscapes of aggression in the hope of grounding their territorial origins.

In *Il secondo libro di prospettiva* (1540), Sebastiano Serlio projected three stage sets for yet another urban theater: the tragic (rational), the comic (circumstantial), and the satyric (bestial). The first two are familiar models, but in the latter is revealed the instrumental not objective precondition for the construction of the city. Here satyrs camp out beneath a canopy of trees: first, to clear the forest and to level the ground, and second, to stockpile timber and to quarry building stone. Sabine Frommel (*Serlio*, Electra, 1998) attributes to Serlio the *Nativita* painting of the same era, revealing a similar didactic setting and cast of characters. Serlio posits similar covenants of source materials for the construction of the profane city as well as the sacred space of a new world occupied by both man and beast.

The nomadic condition of the late twentieth century and the satyric dimension of the new world have been the concerns of my work, my practice as pedagogy for almost four decades now. Surveyors are enlisted in the preparation of the ground, and with the magic of lunatics, all three prepare a sequence of inaugural events. This chorus of the heroes of humanism is the most fundamental precondition for the construction of the city and touches on the collaborative efforts of those who build as well as teach the foundations of our discipline.

My work today is clearly focused on the coincidence of the enduring humanist agenda of "the word made flesh" with Semperian "Specifications for Construction" as both poetic friction and transformative act for here and now. Texts from the humanities are my models of context and strategy: *Genesis* and *The Aeneid* still seem useful starts. New-world explorations of *Robinson Crusoe* and *Walden* prove to be common-sense primers, reading "architecture as a covenant with the world, again." My practice and pedagogy have been founded on the invention of the Spatial Tales of Origin, beginning with Genesis and Exodus, on to Numbers and Acts, and returning to another beginning in recent years: Approximating Stonehenge.

Genesis and Exodus:

> At the scale of the garden before and after the dwelling: The Parasol House revisits the precondition of Eden before traveling on to Jerusalem (1981);

Numbers and Acts:

> At the scale of radical displacement; Noah's Ark and the world re-cited: Parcel X, North Garden, Virginia (1995);

Approximating Stonehenge:

> At the scale of citizens and strangers, approximating north in the public realm: Eric Goodwin Memorial Passage, University of Virginia, Charlottesville, Virginia (2004).

These three projects all witness distinct generations assessing strategies for the construction of site as incremental and ongoing. The notion of "architecture as a covenant with the world, again" is an explicit ethical commitment to the humanities as resonant culture to be sung aloud for every new generation, echoing equally in the Ise Shrine as well as the smoldering ashes of the Vestal Virgins in Eternal Rome.

The word made flesh

Genesis and Exodus

At the scale of the garden before and after the dwelling: The Parasol House revisits the precondition of Eden before traveling on to Jerusalem.

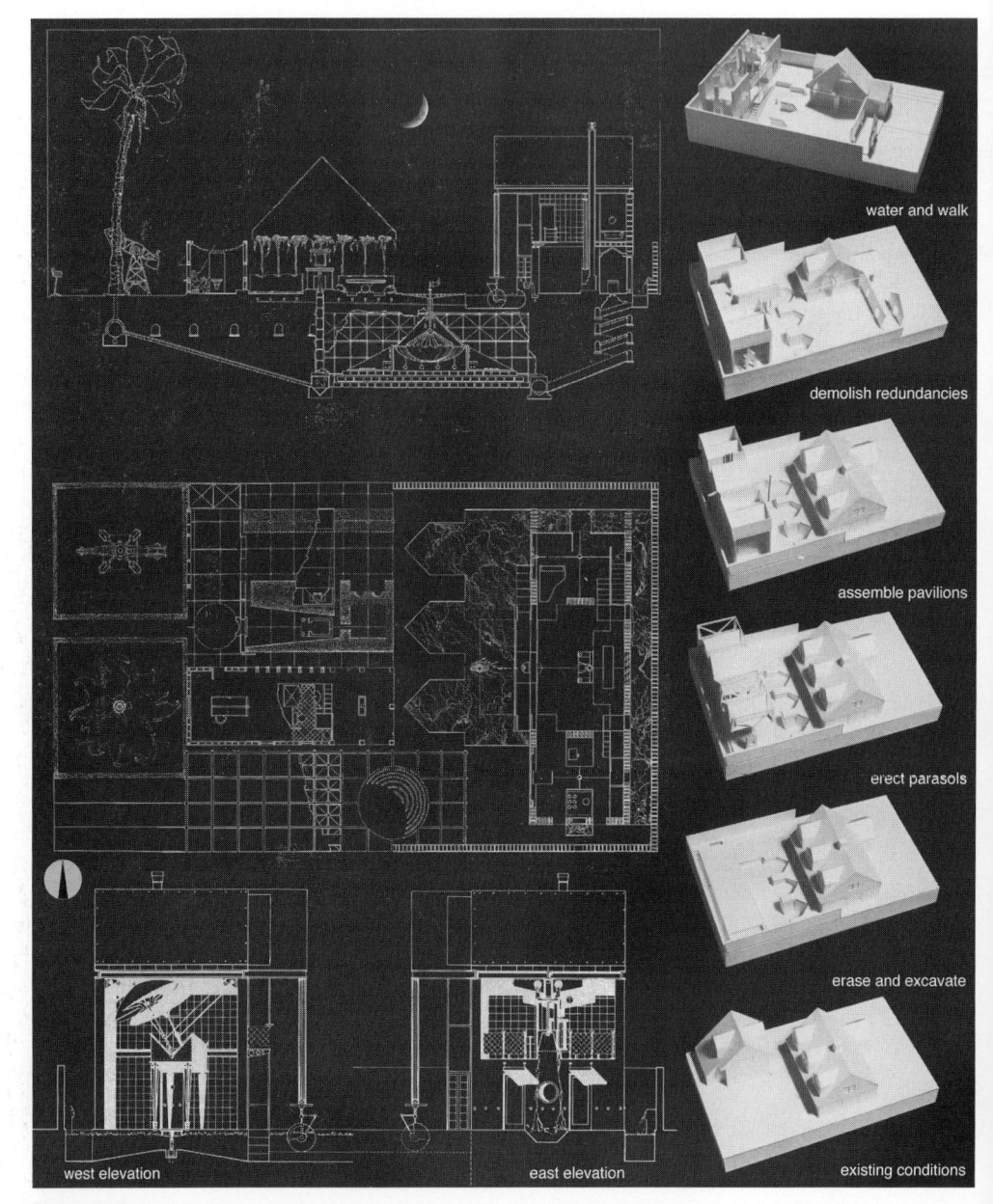

Figure 10.3
Parasol House, Houston, Texas. Plan Section.
Peter Waldman and Chris Genik, 1983

The Parasol House, Houston, Texas 1981/1983: "A tale of two gardens" and the challenge of Serlio's satyric setting

Alice in Wonderland provides us with a narrative of seeing both fantasy and familiarity in a world without scale or gravity. The story is both liberating for a summer's slumbering imagination and terrifying when one awakens to find oneself still on the ground beneath the cooling shade of a tree. Within our landscapes, time never seems to stop, but history is made anew ever so frequently with the invention of pedagogic looking glasses. The Parasol House is a speculation on the necessity of shade in the prismatic light of late twentieth-century urbanism in Arcadian America. The project recalls Boccaccio's *Decameron* where flight from urban chaos to a villa's emancipation transforms nature from wilderness into generative landscapes and accessible gardens.

Professor Seymour Hinge, his Peruvian wife, and their two Izod-tattooed daughters[2] have recently moved from a large prairie-style home in Cincinnati to a small Cape Cod in the heart of Houston. The Hinges have asked their architect to develop a strategy for the transformation of their New England house type into a reconsidered encampment that is responsive to the climate, flora and fauna of hot, humid Houston. The genesis of the proposition corresponds to the progressive stages of an implementation schedule: the back yard turns into a dwelling; while the front of the house turns into an oasis for the dog who is the only resident alien, full-time on-site, rain or shine. For these nomads, one from the north and the other from the south, together with their hybrid children, Houston shimmers as a mirage of a subtropical paradise with terrains gushing black gold beneath a majestic date palm within easy reach out front and a vacated gumbo clay pit out back in need of shade.

Eleven years later, when they leave for Virginia, endlessly in search of Arcadia-in-America, their relationship to nature through the lens of this curious city would be haunted by a resonant space between Genesis and Exodus, between Troy and Rome, between Crusoe's isle and Thoreau's cabin.

The first project (1981) was a plan strategy of inverted subtractions as "specifications" articulating the character of the figures without specifying the spatial preconditions of the field:

1. Existing condition: Cape Cod and servant quarters in hot and humid Houston.
2. Erase the derelict servant quarters and storm-struck remnant of a live oak at the rear of the lot.
3. Excavate a lap pool as the prismatic reflection of the former front dwelling to hyphenate the site.
4. Erect three temporary parasols south of the pool to provide shade from the high noon sun.
5. Assemble three pavilions beneath this shade for a day house below and a night house above.

6. Demolish redundant spaces in the original house leaving one portion as a studio or gatehouse.
7. Water this lawn with morning mist; unpack household gods[3] and walk to work.

The Hinges wanted one spacious place in which to dwell on the ground and one bedroom each for the rising and the setting of the sun, with two baths and a bridge between them, on the second floor. Library, guest quarters and studio in the fragmented section of the former house will provide day-to-day relief, oriented perpendicular to the south-facing pool and new, north-facing parasol house. No mention was made at that time to the character of the dog run.

In the second project (1983), a course change, or revision, became evident when the section of the thin topsoil was mined to yield mirrors for the moon.

Postscript: "A tale of two gardens"

Two years have passed since this project was conceived, and a reconsideration of the landscape has taken place. The previous conception projected two abstract voids between buildings: a lawn in the sun, and a pool in the shade, both connected by a metered yet scaleless sea of concrete. The new landscapes are now particular, not generic, intimate and monumental, rather than scaleless, and are the stuff of fairy tales rather than diagrams. The following "tale of two gardens" is related to family friends by the Hinge children to explain the efforts of their architect, now turned miner of mica, and the influence of his young student on all of them.

Once upon a time, a puppy[4] came to stay with the Hinge family. With family members away at school and work all day, the poor puppy had no alternative but to wander in concentric circles at various staked places in the garden. The intense heat of this concrete ground and of the ever so neatly trimmed lawn was enough to cause extreme dehydration in late spring, followed by debilitating drenching during the summer monsoons. The surrogate day dweller, a dog, would whimper under the hot, unrelenting sun.

Hearing anxious cries one day, a concerned architectural student[5] insisted that the absent-minded professor provide an alternative strategy for the landscape void that had been projected. The student insisted: "Trees must be planted, not abstractly to make a wall in plan, but generously to provide a roof in section. Beneath this roof and in a clearing, a doghouse must be built to guard the former vestibule with an eye kept on fortuitous scraps left on the dining patio to its rear. The former plan of the house must be recalled in the anchoring hearth which will, with time, become the solemn portal to the subterranean pet mausoleum."[6]

Praising shadows, a landscape was projected for the life cycle of the surrogate day dweller, a dog, in place of the empty void made to satisfy the measures of men. The dog has been given not a tethered stake, but "free run" of the former plan. Gathering on occasion by the former hearth or in the shade of the old kitchen porch, dog and family alike settle down to share a meal within the hedges.

While one landscape for the daytime has been cultivated above ground for the dog, family and friends, another more turbulent one has been constructed in a grotto, fashioned by the Izod alligator below grade. A cistern is constructed to catch the acid rain produced from nearby refineries: this water is neutralized by a constant addition of calcium carbonate as base from infertile alligator eggs. This neutralized ground water is then pumped up during the cool of the early morning to refresh the dog's house, which is indeed the garden.

With a grin and a giggle, the Hinge children turn full circle and point to the palm tree and oil pump, proudly displayed on their front lawn to complete this vision of paradise and plenty in a land without pretense and a garden without guilt.

The word made flesh

Numbers and Acts

1/2/4/16/256
Numbers recounts spatial encampments in the wilderness. Old Testament.
Acts constructs the labors of Peter as a foundation builder. New Testament.

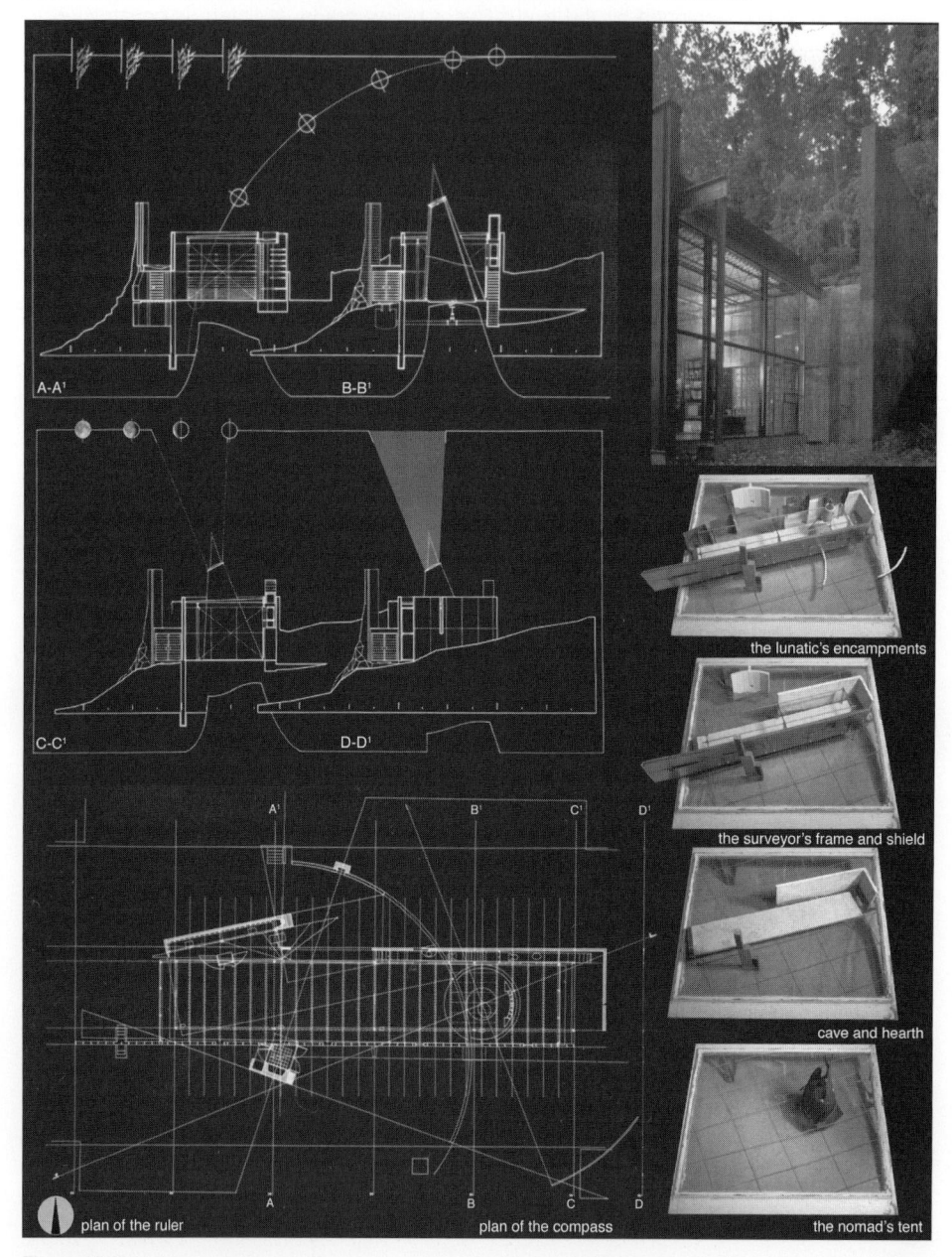

Figure 10.4
Plan X, North Garden, Virginia

Parcel X: North Garden, Virginia 1994
On landscapes within and without

"Architecture as a covenant with the world, again" is a particularly North American preoccupation regarding the cultivation of barren ground without a viable root system upon which to graft new life. Noah's Ark is a spatial and temporal metaphor for starting out afresh with a myriad of collaborators making sure to have two of each kind. The role of the humanities is such an indispensable library for my work and teaching. These are the preconditions of the site offered by the surveyor to Professor Seymour Hinge, en route to Virginia after 11 years in Texas:

> A parcel is a modest but actual fragment of a much larger, if not pretentious, fiction.
> Parcel X, a 3.84-acre remnant, sits at the margin of a still vast pastoral landscape.
> Parcel X is an abandoned site. Too steep for agricultural use, dominated by an ancient poplar forest of little commercial value, only the granite fissures have proffered a crystalline cistern for the Long Arm Valley.

Parcel X was already marked, long before map-makers, surveyors, or soil samplers ever came to project their scars upon this surface. Geological origins were substantially recorded in massive, oblique granite ridges, rhythmically cracked by palisades of virgin poplar shafts. A second slender dimension of understory dogwoods blurs the zone where tall trunks emerge from granite ravines. These ravines are the source of one of the most generous wells in Albemarle County, fed by a constellation of cisterns, slowly being revealed within and without.

This is the Spatial Tale of Origin, inspired by the Incan settlement of Arequipa,[7] signifying a remarkable place to pause in the midst of a journey from the mountains to the sea.

On the first move and the last

To this parcel have journeyed two elderly nomads, now estranged from other vagrant generations, yet faithfully accompanied by their now blind and arthritic dog. Keenly aware of a temporal agenda, they seek to prepare the ground, to tend a garden and eventually to engage the earth.

This project is part of a generational study of climatic dwellings commenced with the Parasol House for the same clients in Houston more than a decade previously. These two campers arrive, one with a ruler, the other with a compass.

On the first day, a tent is pitched, not far from the pre-existing well. A campfire establishes the ash traces of man's first nightmare.

On the second morning, the process of clearing the site provides a staging area for construction. The eastern boundary is the first to be surveyed; a prism pole is left to frame the sun, and the first partial palisade is built, braced against the cold north wind.

At noon a plinth is extended along the full southern edge of the site with a cut and fill operation. That evening, at the western boundary, X-bracing records the setting of the sun in the brittle surface of this first parterre.

Between the palisade and the plinth, the now weary nomads rest under a full moon.

On the third morning, a steel framework is erected, based on a meter of 26 feet to give an enigmatic measure to this now cleared and level site. A fireplace is erected to the south, and a totem appears somewhere beyond the precinct to the north.

On the fourth day, a shield is applied, step by step, to challenge the southern exposure. Steel studs with standing seam copper siding, modulated by Hope's doors, all combine as an incessant template. Thereupon, armatures of eye hooks and guide wires with minds of their own collaborate with Virginia-creeper vines to mask this pretentious straight edge.

On the fifth day, a trailer arrives with kitchen stuff and household gods.

An ark is set upon the ground beneath this masque of urban decorum to serve, temporarily, as a cookhouse/canteen. Similar pairs of packing crates get located more discretely on the uphill side as outhouses, framing a washroom in between. Under the hot noon sun, a sombrero serves as a mirror for the moon, hovering above the previous parterre. From this new terrace one can recover the horizon previously denied by this undulating topography. Late in the day, glass curtain walls seal off the east and the west with the unsentimental anonymity of a two-meter grid. A glass block panel of equal composite size makes prismatic the northern exposure with the hill, bouncing off light from the south.

On the sixth day, a manhole reveals the secrets of a cistern in the bowels of the excavated parterre; beneath a labyrinth of packing crates, a study is perched above with a volcanic lens pointing south, and the ground begins to heave.

That evening, the elderly collaborators find rest in a hammock suspended within this armature of the first campsite and dream now how similar the first move was to the last.

On the seventh day, it is rumored, a wall rises to the north where the totemic stake once distinguished within from without. Its iridescent face now contains an extended aperture, some say barbecue, while others whisper funeral pyre. Only the blind arthritic dog knows for sure the destination of this portal.

In the nomadic North American condition, one can never tell if your next move is to be your last. Precautions should be taken to secure daydreams and nightmares; ancient flues must guard deep cisterns; household goods should be kept at a distance while the preconditions of the site punctuate this campsite from within. This genesis of revelation is the ancient rite of all nomads who know that the city and the garden have origins in the oasis.

The word made flesh

Approximating Stonehenge

Figure 10.5
Goodwin
Memorial
Passage,
University of
Virginia

On specifications for construction: North Porch, Campbell Hall
On the collaboration of allied concrete and a swarm of spiders
On the utility and transformative qualities of tilt slab construction
On sequential markings of the memorable horizon and the totemic vertex
On the strategic responsibilities of surveyors, nomads, and occasional lunatics

113

Eric Goodwin Memorial Passage at the University of Virginia 2004

A remarkable architecture student passed away suddenly before commencement exercises; his peers rose up to erect a memorial at the thickened edge between architecture and landscape. These resultant outdoor classrooms were a student/faculty design build operation to be used routinely to mark time for seminars and studios and to celebrate the annual Ritual of Commencement.

Our site, the north face of Carr's Hill, might not be easily described as level ground. A review of the archive of maps of Carr's Hill since Jefferson's inception reveals numerous fictions as to the location of true north. Thus with Eric's passing and his peers scattering to the winds shortly thereafter, we citizens of Campbell Hall who remained were not certain where north might be precisely located.

We needed to use the strategies of construction to orient ourselves. That summer, we proposed to construct, one by one, with each new moon, concrete slabs determined by the meter of the structure of Campbell Hall. The first slab became the formwork for the next, as one proceeds from east to west. With each sequential pour, the previous one was tilted to the sky; then each totem, braced. Shadows danced, implying a village of teepees.

First, we used both enduring concrete markers as well as ephemeral strings to give measure to this difficult topographic condition.

Second, we used the same fixed points as concrete benchmarks, tables in the broadest sense, together with a system of dynamic approximations to help citizens and strangers alike to find true north as the one requirement of anticipated orientation with this world.

The ethical responsibility of architecture as orientation is the first and only lesson of this north porch.

Third, nomads were to project experimental theaters and landscapes for a tent, a table or two, and a myriad of commemorative and transformative tablets at the scales of both bricks as well as civic mirages.

The spatial setting of the north porch was to be nothing less than the construction site of the intersecting lessons of civic literacy commencing with the ABCs of the Acropolis, onto Bilbao, then the Campidoglio ... with the Ise Shrine as pivotal ... and ending, no doubt, in Zurich at the threshold of a tent perched between the Mountain and the Zee.

Fourth, surveyors constructed concrete markers to measure the horizon first from ground to mountain ridge.

Fifth, lunatics provide upon these foundation plinths additional pours of progressive dimensions now to give measure to the hill as they are then tilted vertically to frame a window to the sky.

Sixth, with time, the tilt slab panels will be incised with the names of departed students and faculty, generous donors and legendary caretakers alike as a prerequisite of citizenship.

Seventh, upon these window plinths a swarm of spiders insert telescoping poles and cables as stanchions for the eventful tent, reliably erected by a band of meandering nomads in the midst of May.

Eighth, it is rumored that another lunatic in the ruins of an ancient fraternity site has supervised a deep casting pit that is quarried as formwork for incubating groundhogs to sustain the stress of tent-induced wind loads.

Ninth, a forest of pylons and correspondent water runnels syncopate the hill.

Tenth, fires burn.

Eleventh, columns begin to dance.[8]

Connective tissue, landscapes of aggression

All these projects are ancient and familiar tales, which have served as an architectural primer for students/speculators, all recounting the enduring codes and components of our discipline. They share a connective tissue demonstrating a syntax of structure all too often forgotten by many in the current amnesia where nothing endures for the contemporary. The preconditions record sites already full, not empty, of geological fissures and resistive soils, where ancient forests are metered by pastoral fences and punctuated by camping sites of nomadic origins.

"Building" as a verb, as an ongoing, phenomenal process, is in crisis, if one also accepts the notion of substantial completion with the assemblage of a checklist, with the assumption that structures are invariant and thus should not creak or leak. These texts are alternatives to an impoverished and pretentious architecture that conventionally values more the resolution, as the stabilization of structure, over the vitality of stress scars and watermarks. Herein is an argument for an architecture that celebrates the instrumentality of "the word made flesh" in landscapes of progressive aggressions. Spatial Tales of Origin recounted in Specifications for Construction should begin and end with yet another eschatological beginning, always found in water and watermarks, soil and stain, in darkness and encrusted patina, in fire and in ash, in secret springs as well as manholes and finally, lightning rods. These projects serve as an apology for the temporary encampment of those in search of a new-world Arcadia where steel frames audibly creak in the wind, concrete displays stress cracks seasonally, and rain shields do not leak (too much).

I have come full circle now, from the child who hit the ground to see the sunlight in dust so long ago in suspended space in New York City, to one now delighting in haunted sites not quite, not totally out of mind. The Parasol House fleshes out a familiar garden in a strange city; Parcel X reveals a nomadic encampment in North Garden, metered by the routines of daily life; and the Goodwin Passage is the secret oasis of the next generation, yet another collaboration with young, fresh-faced students, lunatics all, perhaps, as we strive to polish mirrors for the moon.

Here and now another site looms close at hand for me, teaching at Mr Jefferson's university, inaugurated by the shadows of the Academical Village. The Lessons of the Lawn now also imprints the projects presented herein[9] as ethical strategies for the collaboration of distinct generations to envision common ground within the spatial field. A statue of the blind Homer faces north to Jefferson's Enlightenment Library; at night under a new moon, the Rotunda illuminates Homer and both see through one another. Grounded in the Enlightenment and out in the wilderness, Jefferson gave flesh and marrow to the humanities as useful knowledge in this his last architectural project.

Notes

1. Gottfried Semper, *Der Stil in den technischens und tektonischen Kunsten oder praktische Aesthethik*, cited in William Alexander McClung, *The Architecture of Paradise: Literature Between Eden and Jerusalem*, Berkley, CA: University of California Press, 1983.
2. Peter D. Waldman, "On Landscape: The Parasol House, Houston Texas", *Princeton Architectural Press*, Vol. 3, 1983.
3. Walter Lippmann, "Barren Ground", in *A Preface to Morals*, New York: Brown, Little Publishers, 1929. pp. 61–63.
4. Homer, *The Odyssey*: Ulysses' dog, Argus, remains loyal, waiting for his return to Ithaca.
5. Christopher Genik, former Rice University student, initiated the recurrent dualities referenced to the long-lost natural conditions of the forest and secret spring reconstructed for the Acropolis of Athens in the plan of the Parthenon and the section of the Erectheum.
6. Refers to the tale of the Minotaur, Knossos, and the persistent trope of the labyrinth.
7. Arequipa, Peru, is the law of the Indies city founded by Pizarro in 1530 and where the author served as a Peace Corps volunteer architect from 1967–1969.
8. Joseph Rykwert, *The Dancing Column: On Order in Architecture*, Cambridge, MA: The MIT Press, 1996.
9. Architecture 101 Lessons of the Lawn: "Architecture as a Covenant with the World, Again" is the foundation course on architecture offered to all undergraduates as a foundation course to the humanities by the author since returning from Rome in 2000.

Part 3

Part 3

Measures of awareness

The theme of Part 3 considers the issue of the various modalities of "measure" in the understanding of architectural space, and how these have historically been informed by the humanities. By measure, the authors are referring not to instrumental or abstract forms of calibration, characteristic of modern science, but more ambiguously to its representational understanding, mediated through the *studia humanitatis*. The visual and metaphysical horizons of enquiry that formed the basis of early modern science and epistemology were long endowed with meanings drawn from ancient cosmological symbolism. These relationships, moreover, stand in direct opposition to the modern situation of the apparent disjunction between the humanities and science/technology. By focusing on historical examples, the chapters highlight how scientific and mathematical notions of order, as expressed through geometry, optics, architectural drawing and related concepts of *disegno*, were indelibly imprinted with philosophical and theological principles that gave meaning and purpose to architectural ideas. Questions therefore of "measure" were endowed with symbolic and allegorical content that reinforced an embodied view of the world.

In the first chapter, "Creative inspirations or intellectual impasses? Reflections on relationships between architecture and the humanities", Nader El-Bizri examines developments in medieval optics and geometry – in particular the "geometrisation of place" in the thought of Alhazen. He highlights how early scientific investigations – and their correlative philosophical enquiries – provided essential terms of reference in the *studia humanitatis*. At the same time these investigations contributed towards a deepening understanding of architecture as an embodiment of an ordered cosmology, only later to be redefined as an "applied science" through the growing dominance of technology. The author takes the ideas of Alhazen as an important starting point in this shifting relationship, arguing that the advent of modern concepts of space (in Renaissance theories of pictorial space) were anticipated by Greco-Arabic studies in geometry and optics. This historical background provides a critical point of reference in El-Bizri's assessment of the role of the humanities in architecture today. In spite of the perceived obstacle of the humanities in design creativity, arising from the "burden" of theoretical ideas on design development, El-Bizri suggests that the dangers of "conceptual disappearance" must be brought into the context of the historically rich and creative dialogue between early science and the *studia humanitatis*. Only by understanding this background can we truly understand the value of the humanities in our scientific and technological age.

In Chapter 12, entitled "The Human mind and design creativity: Leon Battista Alberti and *lineamenta*", Nikolaos-Ion Terzoglou investigates the influence of the humanities on the process of creative design. He examines this through an historical and hermeneutical analysis of Leon Battista Alberti's architectural thinking. Focusing on his major work *De Re Aedificatoria*, the chapter provides a new understanding of the relationships between the concept of *lineamentis*, the design process and built architecture. Terzoglou argues that this renewed interpretation can be achieved through the analysis of the possible humanistic and philosophical dimensions of the term *lineamenta*. In this interpretation, Nicholas of Cusa's views on the human mind and its creative nature serve as an important reference. Cusa's idea of a coherent relation between architectural conception, architectural design and architectural praxis is revealed in direct relation to Alberti's humanistic concepts and ideas. The aim of the study is to relocate architectural representation within a broader cultural context that will in turn provide contemporary design processes with an appropriate historical and theoretical dimension.

In Chapter 13, "Renaissance visual thinking: Architectural representation as a medium to contemplate 'real appearance'", Federica Goffi-Hamilton considers the idea of "real appearance" in architectural representation through a detailed examination of a hand-drawn plan of St. Peter's Basilica in Rome by Tiberio Alfarano (1571). The author argues that architectural drawing was understood during the Renaissance not just as a "representation of likeness" but more specifically as "an epiphany of presence". This finds expression in Alfarano's drawing as synonymously building (Basilica), body (St Peter's) and testament (Word of God). Through a detailed examination of the work, Goffi-Hamilton demonstrates how Alfarano sought to convey continuity between Old and New Basilicas, in which the temporal – past and present – and the eternal are intertwined. Considering the multiple meanings of Alfarano's drawing, that transcend singular use, the work raises intriguing questions about the nature and meaning of architectural drawing in our age of digital media.

In Chapter 14, entitled "Neoplatonism at the Accademia di San Luca in Rome", John Hendrix examines the development of the concept of *disegno* at the Accademia di San Luca in Rome, illustrating the continuing importance of Renaissance humanism in the culture of baroque Rome. Hendrix starts by referring to the treatise *L'Idea de' pittori, scultori ed architetti* by Federico Zuccari in which he highlights the Platonic "idea" and its relation to the mind of the artist. Through an examination of various textual sources, from Aristotle's *Metaphysics* to Pietro da Cortona's *Trattato della Pittura e Scultura*, Hendrix explains how the concept of *disegno* and its related classical treatments of cognition were reinterpreted through the ages and became a key theoretical principle in baroque architecture. The author refers to Santi Luca e Martina and San Carlo alle Quattro Fontane as prime examples.

In the final chapter, "Who's on first?", Donald Kunze explores the idea of human subjectivity as an "optical identity", drawing upon the ideas of two thinkers, one at the beginning of the tradition critical of the Enlightenment's positivism (Giambattista Vico), the other near the end (Jacques Lacan). Kunze provides an intriguing comparison between Lacan's concept of "the mirror stage", and Vico's treatment of history which is represented iconographically in the accompanying illustrations of his major philosophical work, *The New Science*. Although these two versions of subjectivity share an uncanny resemblance, Kunze argues that no scholar has yet fully realised their connection in terms that could contribute to a productive basis of an "architecture-as-humanity". The comparison between Vico and Lacan is instructive to contemporary architectural discourse as it demonstrates how metonymic thought can encompass the dual roles of theoretical enquiry; at once "coherent" in its unity and "contingent" in its fragmentation.

Chapter 11

Creative inspirations or intellectual impasses?

Reflections on relationships between architecture and the humanities

Nader El-Bizri

> Il y a un sens à prendre la maison comme un *instrument d'analyse* pour l'âme humaine.
>
> (Gaston Bachelard, *La poétique de l'espace*)[1]

Figure 11.1
Between the lines

Dialectics of architecture and the humanities

It is customarily argued that the various disciplines in the humanities present selected intellectual directives that could potentially inspire the unfolding of creative architectural design processes, or orient their conceptual expressions, their imaginative projections and concretised representations. This state of affairs is commonly supported by series of discourses, which are typically articulated around historical precedents that co-entangled the history of architecture with the history of the exact sciences, of mathematics and philosophy.

The notions that are derived from the humanities may assist architectural thinking and its intellective ponderings within the diverse spheres of the *production* of architecture and the critical academic study of their design determinants. It is ordinarily believed that the humanities historically cultivated the didactic refinement of the character of the architect, in addition to inspiring pursuits of excellence in design, and the thoughtful reflection on their actualised consequences in accommodating evolving material cultures within the sequence of civilisation.

The humanities stimulated the generation of diverse measures of awareness with respect to the differential historical, political, social-cultural and economic factors, which dialectically mould the situational conditions of design.[2] Nevertheless, the disciplinary quarrels between competing schools of thought across the humanities, with their disputes over methodology and epistemic measures, do at times result in intellectual disorientations. This leads to conceptual situations that are marked by suspended or deferred judgements, which may also seem to be groundless in terms of their veridical criteria, or the "truth value" of their theoretical findings and utterances. Nevertheless, the humanities do in most instances offer multiple conceptual frameworks that assist enquiring minds in terms of grounding their analytic skills and critical methods, in addition to deepening the bearings of their acquired knowledge. Moreover, the humanities grant some concepts that facilitate the efforts to intertwine selected professional practices in architecture with critical reflections on human comportments that pertain to the plural modalities of exercising societal justice, or establishing political-legal procedures that sustain civic orders, along with their associated ethical, communicative and symbolic orientations.

Although the letters, the arts and sciences of the humanities furnish potential inspirational and informative directives with respect to the unfurling of design in its material-formal qualities, they may also (under certain contemplative, critical and analytical circumstances) arrest the free flow of their creative impetus, or encumber it with conceptual burdens, which accommodate the gradual or abrupt dominion of the discursive and the textual over the intuitively suggestive potencies of imagery (perhaps in expression of the long-standing tension between *scripture* and *iconography*). The *philosophème* (as philosophical-logical enunciation) may thus be granted an unquestioned station of conceptual primacy over the *graphème* (as graphical trace) with its relatable potential material manifestations that get projected onto concretised built forms. The theories of the humanities and their methodologies may still conflict with architectural thinking, which does not always necessitate the same level of academic resistance to the temptations of adopting facile conceptual formulations of eclecticism or syncretism, while also eschewing prolonged textual enquiry. The controversies of scholarship in the humanities, along with their unresolved or renewed debates over methodology, with their posited anomalies or paradoxes, all may contribute to impeding the liberality of creativeness in design and its penchant to be selective, or sometimes to be fragmentary in identifying notions that inspire its expressions or serve its representations; and doing so without methodological strictures or logically formalised structures of argumentation, discursive reasoning and inferential explications.

"Architectural humanities" seem to oscillate between coherent, orderable and directional conceptual juxtapositions from one side (which advance convincing lines of sound reasoning) and, from the other side, selections of concepts that are chosen randomly, or nominally, and grouped together by way of naive analogisms or in terms of impulsive inspirations, which nonetheless serve design ends and project *meaning* unto the technological elements of architecture as "applied science". A deepened relationship of theoretical dependency on the humanities may possibly preoccupy the design endeavour with methodological/technical quarrels (contemporary, or deeply entrenched and inherited), which do not necessarily serve the purposes of architecture with its evolving situational finalities and the manner it integrates technology in "architectural engineering".

The dialectics of the interactions of architecture with the humanities evolved historically in terms of responses to developments in philosophical and scientific ideas, and in reflection of the concrete elements of the unfolding of material cultures and technologies within the inter-cultural lineages of the sequence of civilisation. To better grasp the conceptual catalysts that modulated the relationships of the humanities with architectural history, theory, criticism and design, we ought to account for concretised precedents that are documented through the methods of historiography, which also represented actual instances of the impact of the history of the exact sciences, of mathematics and philosophy, on certain fundamental outlooks in the history of architecture. This line of enquiry encompasses reflections on the classical sources of the historical engagement of the humanities with the "exact and mathematical sciences", which predated "modernity" and the Renaissance, and partly informed their intellectual evolutions.

A foundational scientific legacy

It is believed that "the first plausible anticipation of modernity" is witnessed through the complex circumstances that resulted in the formation of pictorial perspective in the Renaissance.[3] However, the principal "scientific" directives of this *perspectiva* legacy were partly inspired by Greco-Arabic studies in optics, as they were adaptively appropriated and mediated through medieval Latin scholarship. The exact and mathematical sciences had epistemic bearings on rethinking fundamental questions that preoccupied classical philosophers and the *curricula* of medieval scholasticism (including the *quadrivium*). This state of affairs was manifested through reflections on the acquisition of knowledge and the pathways to disclosing "truth", or uncovering "the ultimate principles of reality". These conceptual developments found one of their *foundational* expressions in the scientific legacy of the Arab polymath Alhazen (al-Hasan Ibn al-Haytham; born 965 CE in Basra and died ca. 1041 CE in Cairo); as particularly embodied in his revolutionising reform of the classical science of optics, his "geometrisation of place" and the epistemic influence of his research on Renaissance scholarship.

It is well documented that Alhazen's studies in optics impacted the unfolding of the *perspectiva* traditions in the history of science of thirteenth- and fourteenth-century European scholars; particularly in Franciscan and Dominican workshops and mainly through the efforts of opticians like Roger Bacon (*Opus maius*; *De multiplicatione specierum*), John Peckham (*Perspectiva communis*), Witelo (*Perspectiva*) and Theodoric of Fribourg (*De iride et radialibus impressionibus*).[4]

Alhazen's *Optics* (titled in Arabic as *Kitâb al-manâzir* and composed between 1028 and 1038) was translated into Latin towards the end of the twelfth century by Gerard of Cremona, under the title *Perspectiva* or *De Aspectibus*.[5] An Italian translation of Alhazen's text, titled *Prospettiva*, was also produced in the fourteenth century, and a Latin version was eventually printed in Basel in 1572, under the editorship of Friedrich Risner, entitled *Opticae Thesaurus Alhazeni*. All these renderings of the manuscripts of Alhazen's *Optics* were widely disseminated within European scholarly circles, from the twelfth century till the seventeenth century.[6]

It is believed that the most remarkable "revolution" in the history of classical optics, from the times of Ptolemy to those of Johannes Kepler, is embodied in Alhazen's book of *Optics*.[7] This line in scientific and mathematical research, with its associated corollaries in geometrical, physical, physiological and meteorological optics (as initiated by Alhazen and assimilated by Latin medieval scholars), eventually informed the unfolding of the Renaissance *perspectiva* traditions. This was mainly the case with the theoretical studies of fifteenth-century luminaries like Leon Battista Alberti (*De pictura*) and Lorenzo Ghiberti (*Commentario terzo*), as mediated also by the teachings of Renaissance scholars of the calibre of Biagio Pelacani da Parma (*Quaestionis perspectivae*), who partly founded his natural philosophy on Alhazen's legacy instead of scholasticism.[8] It is even argued that the "perspective experiments" of Filippo Brunelleschi resulted from the revival of Alhazen's scientific tradition in the fourteenth century.[9]

Alhazen established "the experimental method" based on systematisations of the research of ninth and tenth-century Arab polymaths of the Archimedean-Apollonian tradition. He used controlled testing and mathematical structuring in optical experiments, including studies in dioptrics and catoptrics. He also demonstrated how "vision" occurred by way of exercising the faculty of discernment and judgement (*virtus distinctiva*) and argued that visual perception was a reliable source of scientific knowledge in studying natural phenomena and the disclosure of their reality; precisely when it is conditioned by exact experimental observations and mathematical modelling. This epistemic aspect contrasted with the principal intellectual traditions that were prevalent in his age, or those that continued to exercise an influence on the speculations of medieval scholasticism and the metaphysical conjectures of Renaissance thinkers. Unlike most of his contemporaries and successors, he was not a theologian, and he did not study light and vision from onto-theological standpoints. His research neutralised the impact of metaphysics (and "religiosity") in studying natural phenomena and solving mathematical problems. Alhazen was indifferent to the "symbolism" of Neoplatonist and neo-Pythagorean doctrines, and he critically reassessed the Aristotelian notions in physics, based on experimentations and mathematical demonstrations. His scientific legacy accomplished an epistemological shift from focusing on texts (revealed scripture, or classical exegesis and commentary) towards exploring the veridical conditions of visual perception in the acquisition of rational knowledge and the unveiling of truth. His studies already anticipated the main directives of modern science and the epistemic mood of the seventeenth century in deontologising the notions of physics and mathematics, as well as avoiding the "metaphorical" connections that were instated (theologically) between textual interpretation and visual atonement (namely, the measuring of truth and ontological reality by way of a *visio intellectualis*).[10]

The science of optics

Alhazen resolved the long-standing dispute between the ancient mathematicians (Euclidean and Ptolemaic) and the ancient physicists (Aristotelian) over the nature of vision and light in his *Optics*. He demonstrated that vision occurs by way of the introduction of physical light rays into the eye, following the structuring shape of a geometric *virtual* pyramid/cone of vision (*makhrût al-shu'â'*) with its apex at the centre of the eye and its base on the lit visible surfaces of the object of vision. Alhazen rejected the theory of the ancient mathematicians, which entails that vision occurs by way of the *emission* of subtle non-consuming rays of light from the eye that meet the ambient and transparent lit medium and illuminate the visible object, while being physically structured in the form of an *actual* pyramid/cone. He also refuted the account of the ancient physicists, which ambivalently conjectured that vision results from the *intromission* into the eye of the form of the visible object, as abstracted from its matter, when the transparent medium (*diaphanes*) between the observer and the observed is actualised by physical illumination.[11]

Alhazen devised novel methodological procedures that brought the certitude and invariance of geometrical demonstration to bear with isomorphism on his research in physics. He moreover subjected the resultant mathematical-physical models and hypotheses to experimentation (*i'tibâr*) by way of empirical procedures of controlled testing and purpose-built experimental installations, including the camera obscura (*bayt muzlim*). Alhazen studied vision in terms of geometrical models based on the mathematical analysis of the pyramid/cone of vision, and he investigated the structure of the eye in anatomical terms, and showed how the crystalline humour admits *without refraction* only the light rays that are perpendicular to its outer spherical surface. This correlated with his psychological-physiological explication of binocular vision, and his investigation of stereopsis (stereoscopy) and parallax ocular phenomena, along with the examination of the veridical conditions of visual perception in terms of spatial displacement and against the horizon of the flux of time. This corresponded also with his investigation of the comportment of light, its rectilinear propagation within a transparent medium, its refraction when passing across transparent media and its spherical irradiation from self-luminous sources or lit surfaces. He also mathematically modelled the principles and instruments of catoptrics and dioptrics and studied them experimentally.[12] His geometrical, physical, physiological and meteorological studies in optics were related to his psychology of visual perception and to his analysis of the faculties of judgement and discernment (*virtus distinctiva*), of cognitive comparative measure, *eidetic* recognition, imagination and memory.[13]

Prolongations of the *perspectiva* traditions

As a method of pictorial representation, which also assisted the design endeavour in the organisation of architectural space, *perspective* necessitated the devising of instruments to aid visual perception and to reduce the errors of sight or the occurrence of perceptual illusions. Ultimately, the history of Renaissance art and architecture, which was inspired by the history of classical optics, acted in its turn as an active catalyst for the refinement of tools that eventually assisted in expanding the spheres of scientific enquiry. This search for exactitude in the instrumentation of scientific research in the Renaissance, and its related quest for visual and optical certainty, was informed by the equally prevalent emphasis on historical and logical truth in the *studia humanitatis*. This state of affairs signalled the attempts to overcome the focus on the divine *nomos*, in terms of reforming dialectics and deontologising the "grammar of the word" in the determination of truth.[14]

     New mathematical formulations of optics, dioptrics and catoptrics, in the seventeenth century, led to technical advancements in the design and production of finely polished lenses, with better controls over aberrations. These were deployed in novel technologies that were associated with the manufacturing of spectacles and the production of the early refractive and reflective telescopes, along with the mathematical and physical study of the potentials of their optical augmentations through smaller refined concave and convex lenses. By the end of the seventeenth century and the beginning of the eighteenth century further strides were made in optics and its associable lines of specialised instruments that resulted in the production of compound microscopes.[15]

The search for veridical criteria of sound vision in classical optics, primarily since the times of the revolutionising reform of Alhazen, and in terms of the unfolding of prolongations of his *perspectiva* tradition in the history of science, all informed the mathematical and physical research efforts and their applications in astronomy, mechanics, meteorology and the perfection of scientific instruments (astrolabes, sundials, celestial globes, compasses and lenses). This epistemic impetus in the advent of material culture resulted ultimately in ambitious endeavours to visually verify the *truth* of the excessively remote and far (*telescopy* in astronomy, physics and mechanics), as well as the veracity of what is extremely minuscule (*microscopy* in microbiology, embryology, botany and crystallography).

Classical optics informed Renaissance methods of representing spatial depth, and its articulation in "heterodox bifocal perspectives", since the *trecento*.[16] The mathematical-physical investigation of the veridical conditions of vision, along with the endeavour to perfect the cognitive procedures and tools to aid sight in avoiding perceptual fallacies, all reflected an *inherited* epistemic classical distrust of the fallibility of the senses and their predisposition to error. This phenomenon, which is associated with the observations of optics, the demonstrations of mathematics, the investigations of physics and the speculative deliberations of metaphysics, found reinforcing expressions in the concerns of Renaissance theorists and practitioners in art and architecture, who in their turn developed devices to aid them in establishing "legitimate" geometric constructions of perspective. Such optical aids were gradually perfected in terms of their expanded applications within the history of technology and its manifestations in material culture. Classical optics, which inspired architecture and art, was itself served by the search for the veridical conditions of vision and visual representation in the Renaissance; architectural thinking was active in the production of scientific knowledge and the grounding of culture. These historical dynamics cumulatively led to further progress in early modern optics, which was gradually transformed into a techno-science of visual instruments (telescopes, microscopes, perfected spectacles and lenses) that opened up the horizons of modern science.

Geometrising place

The foundational aspects of the impact of the history of the exact sciences, of mathematics and philosophy, on the unfolding of fundamental notions in architecture, were also witnessed with the "geometrisation of place", and the conception of place as a "mathematical extension *qua* space". This historical endeavour in the mathematisation of a fundamental notion in physics leads us again to examine the mathematical legacy of Alhazen as exemplified in this context by his *Discourse on Place* (*Qawl fî al-makân*).[17]

The mathematicians of the ninth- and tenth-centuries' Islamic civilisation expanded the classical scientific traditions of Euclid, Ptolemy, Archimedes and Apollonius of Perga. These emergent new spheres of mathematical research centred on investigating geometrical transformations (similitude, translation, homothety, affinity), the enactment of cinematographic geometric demonstrations, the unprecedented introduction of *motion* (*kinêsis*) as a principle of physics within geometry, in addition to the use of projective methods in spherics, the squaring of curvilinear figures and the study of the anaclastic properties of conic, cylindrical and spherical sections. Novel solutions were devised to ancient geometrical problems and new questions in mathematics were formulated, including the establishment of complex methodologies of mathematical research that applied the newly founded discipline of algebra (systematised in the ninth century by Khwârizmî) to arithmetic and geometry, which resulted in the initiation of novel branches of research in mathematics and their applications in astronomy, mechanics, physics and the construction of scientific instruments.[18]

Besides the need to respond to these breakthroughs in mathematics, Alhazen's "geometrisation of place" was also undertaken in view of grounding his epistemological penchant to offer mathematical solutions to problems in theoretical philosophy. These factors necessitated that he devise a new understanding of place as "space", which in its turn required the rejection of the Aristotelian definition of *topos* in Book *Delta* of the *Physics* (212a: 20–21) as "the innermost primary *surface-boundary* of the containing body that is at rest, and is in contact with the outermost surface of the mobile contained body".

Using mathematical demonstrations that were applied on geometrical solids of equal surface areas and on figures that have equal perimeters, Alhazen showed that the sphere is the largest in (volumetric) size with respect to all other primary solids that have equal surface areas. So, if a given sphere has the same surface area as a given cylinder, they occupy equal places according to Aristotle, and yet, the sphere would have a larger (volumetric) magnitude than the cylinder; hence unequal objects occupy equal places, which is not the case. In refuting the Aristotelian *physical* conception of *topos*, Alhazen posited *al-makân* (place) as an imagined/postulated void (*khalâ' mutakhayyal*) whose existence is secured in the imagination, as is the case with invariable geometrical entities. He moreover held that the "postulated void" *qua* "geometrised place" consisted of imagined immaterial distances that are between the opposite points of the surfaces surrounding it, which get united as isomorphic mathematical *cum* virtual dimensions.

Alhazen's geometrised place resulted from a mathematical isometric "bijection" function between two sets of distances.[19] Consequently, nothing is retained of the properties of a body other than its *extendedness*, which consists of mathematical distances that underlie the geometrical conception of place as a metric *region of extension*. This spatial mathematisation of place was correlative with Alhazen's affirmation of the phenomenal "visibility of spatial depth" in his *Optics* and the visual estimation of the magnitude of visible distances.[20] The phenomenon of the "visibility of space" became fundamental in deliberations over the legitimacy of projective methods in pictorial representation and the geometric construction of perspectives. This ushered debates over orthogonals, viewing points, vanishing points and the visual pyramid/cone, which became foundational concepts for the invention of geometrised linear perspective and its presupposition of a mathematised "looking space". Moreover, the evolution of Euclidean geometry benefited immensely from "the mathematisation of place", which among other developments resulted in the emergence of what came to be known in periods following Alhazen's age as "Euclidean space" (namely, a relatively modern notion that is historically posterior to the geometry of figures as embodied in Euclid's *Elements* and *Data*). Furthermore, Alhazen's eleventh-century "geometrisation of place", with its resultant definition of *makân* as geometrical space *qua* extension, was itself reconfirmed in mathematical-physical terms in the seventeenth century; particularly through Descartes' conception of "*extensio*" and Leibniz's "*analysis situs*".[21]

The question of space

It is convincingly argued that "architecture presents us with the sense of coherent spatiality as embodied in the notion of the *room*" and that "this idealised representation acquired in its long-standing history the characteristics of the isotropic space of geometry", which is manifest in the *perspectival* representation of architectural localities and is also confirmed by the observation of architecture in terms of "the appearing parallelism of its structuring components like columns, pillars and walls, in addition to the axial regularity of its spatial articulations and orderings".[22] Yet, architecture entangles the isotropic and *homogeneous mathematical space* of abstraction with the unique and *heterogeneous physical place* of experiential phenomenal concreteness. Natural *perspective*, as an "optical phenomenon", refers essentially to the *mystery of depth* and its visibility; and in its technical determination, as a construct of visual representation, it expresses the deep "desire for depth" (*un désir de profondeur*) that ultimately integrates the impetus of geometry and optics within the structuring of pictorial perspectives and relatable modes of organising space.[23]

Architectural "space" (*qua* "room") is grasped by way of human situational, topological and corporeal experiences prior to being determined geometrically. The debates over the *costruzione legittima* of perspective rested on the geometry of the virtual pyramid/cone of vision, the disclosure of the vanishing point and the proportional construction of pictorial foreshortenings. However, the *costruzione legittima* constituted a reductive pictorial order that is removed from concrete reality by way of disembodiment and was based on geometrical projective formalisations of visual experience.[24] Behind the geometric construction of a pictorial "illusion" (two-dimensional "semblance") of depth and the mathematical determination of place, exists the phenomenal/situational embodied experiencing of spatial depth and its concretised visibility. These phenomena were carefully studied by Alhazen in his *Optics*, and were echoed by Lorenzo Ghiberti in his *Commentario terzo* and in Leon Battista Alberti's reflections on the *costruzione legittima*.[25]

The seventeenth and eighteenth centuries' focus on the determination of space, in terms of mathematics, physics and mechanics, received various forms of resistance, including the refutation of the visibility of space in the immaterialism of George Berkeley.[26] However, the most poignant account of alternative modes of rethinking space was articulated in Kant's *Critique of Pure Reason* (*Kritik der reinen Vernunft*, A23, B37–A30, B46). Space, which conditions the possibility of experiencing, acts as a prerequisite for affirming the universal applicability of geometry and its synthetic a priori propositions. Space (*Raum*) is not an empirical notion that is derived from outer experience; it is rather "*a pure a priori [subjective] form of outer intuition*" that underlies appearances.

This state of affairs confronts us with the ontological problem of determining "the kind of *being* of space", whether in terms of dominant mathematical-physical models, or through philosophical efforts that account for space by way of subjective embodiment, kinaesthetic ocular-motor constitutions, or the existential-experiential analytics of *being-in-the-world* (*in-der-Welt-sein*) and *lived* spatiality (*l'espace vécu*). This ontological dilemma was expressed in Heidegger's *Sein und Zeit* (section 24);[27] that "the fact that space *shows* itself *in a world* does not tell us anything about its kind of *being*" since "the *being* of space" is not the same kind of *being* (*sein*) as that of the *res extensa* or the *res cogitans*.[28] The decisive task in grasping the ontological problem of space lies in freeing up the question of the being of space from the narrowness of undifferentiated and random concepts of *being*. The attempts to clarify the being of space are moved in the direction of elucidating the question of "the meaning, truth, and place of being"; particularly during an epoch dominated by the unfolding of the essence of technology, and in an era that is posited by some as being that of "the closure of metaphysics".[29]

Horizons and obstructions

The academic study of architecture in terms of orientations and methods derived from the humanities yields its own ontological and epistemological entailments. This calls for a philosophical interrogation of some of the presuppositions that underpin the grasping of the humanities as sources of textual inspiration, of intellectual rigour, of conceptual coherence and methodical study, which inform architectural thinking and dialectically confront its unfolding with challenges regarding the means of communicating design ideas, their theoretical grounding and notional understanding, in addition to orienting the integration of technology in the making of architecture. The restoration of dialogues between intellectual enquiries in the humanities and creativity in design, or the transformation of such potential encounters, invites a contemplation of the impetus and equipoise of discourses in the humanities that affected architectural thought, its assimilation and organisation of spatial emplacements, the technical implementing of their materialised forms and the accommodation of their multiple uses.

"The happening of *truth*" in the *work* of architecture; whether it results from artistic, literary, techno-scientific, philosophical or theological endeavours, each has its nuances, its internal riddles, or "clear and distinct ideas". These present us with occasions for interrogating architecture and its possible dialogue (or *monologue*!) with the humanities. This is the case despite the purported "crisis" of some of the humanities disciplines, or the noticeable current "institutional" bent on marginalising the merits of their revelatory, discursive, didactic, apodictic or pedagogic functions in contemporary *academia*, with the dominance of quantification in learning and research, or the attempted reductive measuring of thought by way of expedient, utilitarian, mercantile and technological scales.

Whether with the historical development of the experimental method and its mathematical structuring, or by way of the geometrisation of place and the notions of physics (as early as the unfolding of Alhazen's legacy), knowledge was gradually disengaged from its cosmological, metaphysical and theological moorings, and the determination of truth shifted in epistemic terms away from the authority of the revealed word. Quantifications, structural formalisations, in addition to deontologising epistemology, all contributed to technological and instrumental appropriations of scientific and mathematical knowledge. The turn towards positivist verifications, experimentation and mathematical modelling, and the shift away from textual directives (revealed, philosophical and literary), undermined the role of the humanities in the disclosure of truth and the evolution of "visual culture". In reflection on this predicament, those who examine the "*archéologie des sciences humaines*" with epistemic seriousness also believe that the object of knowledge and research, the "human being" (*l'homme*) or "humanism", is a recent phenomenon of "enquiry".[30] Like its conceptual appearance is arguably historical and partly resulted from internal mutations in culture, it is potentially destined to disappear (as a "notion") and become the object of archaeology, albeit being contemporary. And yet, if there is indeed some sort of crisis in its potential *conceptual* disappearance, this may ultimately prove to be liberating for novel possibilities in thinking that may potentially lead to intellectual renewals!

Notes

1. Gaston Bachelard, *La poétique de l'espace*, Paris: Presses Universitaires de France, 1957, p. 19.
2. This is expressed in terms of responses to ecology, to the political mediation of power, to the dialectics of instating ideology, the application of technology, or the institutionalisation of deference to patronage and *capital*.
3. Dalibor Vesely, *Architecture in the Age of Divided Representation: The Question of Creativity in the Shadow of Production*, Cambridge, MA: The MIT Press, 2004.
4. We ought to mention Robert Grosseteste in this context; and yet, to signal that he was inspired by the *De Aspectibus* of the ninth-century Arab philosopher al-Kindî (in expression of optical theories attributed to Euclid and Ptolemy), and who was not aware of Alhazen's revolutionising reform of optics. See Alexandre Koyré, *Etudes d'histoire de la pensée scientifique*, Paris: Gallimard, 1973, pp. 69–71; David C. Lindberg, "Lines of Influence in the 13th century Optics: Bacon, Witelo, and Pecham", *Speculum* 1971: 46, pp. 66–83.

5   Several manuscripts of the Latin renditions of Alhazen's *Optics* are extant, and these are attributed to channels of transmission via Andalusia and Sicily.
6   The manuscripts of Alhazen's *Optics* were studied by medieval and Renaissance polymaths, in addition to early-modern scientists like Kepler, Descartes, Huygens and Newton, who consulted the Latin printed 1572 edition of Alhazen's text. See Mustafâ Nazîf, *al-Hasan bin al-Haytham: Buhûthuh wa kushûfuh al-basariyya*, 2 vols. Cairo: Matba'at al-nûrî, 1942–1943; Matthias Schramm, *Ibn al-Haythams Weg zur Physik*, Wiesbaden: Steiner, 1963; Abdelhamid I. Sabra (trans.) *Ibn al-Haytham's Optics, Books I–III: On Direct Vision*, London: Warburg Institute, 1989.
7   Nazîf, Schramm; Sabra; *op. cit.*
8   Refer to: Graziela Federici Vescovini, *Studi sulla prospettiva medievale*, Turin: Giappichelli, 1965, pp. 239–72; Graziela Federici Vescovini, "L'influence du *De aspectibus* (*Kitâb al-Manâzir*) d'Alhazen dans le moyen âge latin", *Archives internationales d'histoire des sciences* 1990: 40, pp. 220–38; Dominique Raynaud, "Une application méconnue des principes de la vision binoculaire: Alhazen et les peintres du *trecento* (1295–1450)", *Oriens-Occidens* 2004: 5, pp. 93–131; Dominique Raynaud, "Le tracé continu des sections coniques à la Renaissance: Applications optico-perspectives, héritage de la tradition mathématique arabe", *Arabic Sciences and Philosophy* 2007: 17, pp. 299–345.
9   Graziela Federici Vescovini, "La prospettiva del Brunelleschi, Alhazen e Biagio Pelacani a Firenze", in *Filippo Brunelleschi: la sua opera e il suo tempo*, Pina Ragionieri (ed.), Florence: Centro Di, 1980, pp. 333–48; David Summers, *Vision, Reflection, and Desire in Western Painting*, Chapel Hill: University of North Carolina Press, 2007, pp. 53–55.
10 For instance, the *metaphysical* propositions of Nicolaus Cusanus in this context would not have accorded with Alhazen's *scientific* research. Moreover, the exploration of the *berillus* by Cusanus carried onto-theological entailments that are not integral to the science of dioptrics. See Nicolaus Cusanus, "De beryllo", in *Opera omnia*, II.1, Ludwig Baur (ed.), Hamburg: F. Meiner, 1940, pp. 4–7; and Vesely, *Architecture in the Age of Divided Representation, op. cit.*, pp. 160–63.
11 I investigated this matter elsewhere: Nader El-Bizri, "A Philosophical Perspective on Alhazen's *Optics*", *Arabic Sciences and Philosophy* 2005: 15, pp. 189–218.
12 Alhazen studied the reflection of light on variegated polished surfaces, and the refraction of light when passing through transparent media that differ in terms of their refractive indexes, along with the respective instruments of catoptrics and dioptrics, as curved or straight mirrors and lenses.
13 El-Bizri, "A Philosophical Perspective on Alhazen's *Optics*", *art. cit.*
14 This is partly highlighted in the historiography of Lorenzo Valla (*Repastinatio dialectice et philosophie; Elegantiae linguae Latinae*) and in Alberti's optical studies (*De pictura*). See Manfredo Tafuri, *Interpreting the Renaissance. Princes, Cities, Architects*, Daniel Sherer (trans.), New Haven, CT: Yale University Press, 2006, pp. 48–50.
15 To evoke the research of Robert Hooke and Antonie van Leeuwenhoek.
16 Raynaud, "Une application méconnue des principes de la vision binoculaire", *art. cit.*
17 Roshdi Rashed, *Les mathématiques infinitésimales* IV, London: al-Furqân Islamic Heritage Foundation, 2002, pp. 666–85.
18 I investigated this matter elsewhere: Nader El-Bizri, "In Defence of the Sovereignty of Philosophy: al-Baghdâdî's Critique of Ibn al-Haytham's Geometrisation of Place", *Arabic Sciences and Philosophy* 2007: 17, pp. 57–80.
19 This designates a combination of an "injection" *qua* "one-to-one correspondence" and a "surjection" *qua* "onto" function.
20 I investigated this matter elsewhere: Nader El-Bizri, "La perception de la profondeur: Alhazen, Berkeley et Merleau-Ponty", *Oriens-Occidens* 2004: 5, pp. 171–84.
21 El-Bizri, "In Defence of the Sovereignty of Philosophy", *art. cit.*, pp. 67–69.
22 Vesely, *Architecture in the Age of Divided Representation, op. cit.*, pp. 113, 140–41.
23 I investigated this matter elsewhere: Nader El-Bizri, "A Phenomenological Account of the Ontological Problem of Space", *Existentia Meletai-Sophias* 2002: 12, pp. 345–64.
24 Vesely, *Architecture in the Age of Divided Representation, op. cit.*, pp. 148–49.
25 *Ibid.*, pp. 163–65.
26 El-Bizri, "La perception de la profondeur", *art. cit.*

27 Martin Heidegger, *Sein und Zeit, Gesamtausgabe Band 2*, Frankfurt am Main: Vittorio Klostermann, 1977.

28 Based on this, space is not reducible to a geometrical *extensio*, as Descartes proclaimed, or to an objective absolute, like Newton argued, or to a relational quantifiable function, as Leibniz conjectured. Moreover, one would doubt the sufficiency of Kant's positing of space as a pure a priori subjective form of outer intuition; and, in a critical turn one would also question Edmund Husserl's claim that space is constituted by transcendental subjectivity in terms of kinaesthetic corporeal functions. I examined this question elsewhere: El-Bizri, "A Phenomenological Account of the Ontological Problem of Space", *art. cit.*

29 I investigated this matter elsewhere: Nader El-Bizri, "*ON KAI KHÔRA*: Situating Heidegger between the *Sophist* and the *Timaeus*", *Studia Phaenomenologica* 2004: 4, pp. 73–98.

30 Michel Foucault, *Les mots et les choses: une archéologie des sciences humaines*, Paris: Gallimard, 1966.

Chapter 12

The human mind and design creativity

Leon Battista Alberti and *lineamenta*

Nikolaos-Ion Terzoglou

Architectural creation and the humanities

The present article investigates the influence of the humanities on the process of creative design through a historical and hermeneutical analysis of Leon Battista Alberti's architectural thinking. Focusing on his major work "*De Re Aedificatoria*", the article attempts to provide a new understanding of the relationships between the concept of *lineamentis*, the design process and built architectural space.[1] It is argued that such an interpretation can be achieved through the analysis of the possible humanistic, philosophical, social and cultural meaning of the term "*lineamenta*". The aim of the above elaborations is to relocate architectural representations in a broader cultural context and to restore the seemingly forgotten dialogue between intellectual inquiry in the humanities and design creativity, thus enabling contemporary architectural design processes to acquire a critical and social dimension. It is asserted that a real contribution of the recently established *digital representational media* of architectural design to the humanistic quality and the ideational content of contemporary architectural thought and praxis should involve a critical analysis of their historical, logical, social, cultural and philosophical presuppositions.[2]

Leon Battista Alberti and the concept of *lineamentis*

The text "*De Re Aedificatoria*", written by Leon Battista Alberti around 1450, can be considered as the inauguration of the modern conception of architecture. This conception understands architecture as a mental and spiritual creation, thus reflecting the new cultural ideal of Renaissance life as a unified synthesis of thought and praxis and the relevant social demand for an enlarged role of the individual into the civic government of the independent city states of fifteenth-century Italy.[3] In the Introduction of the work, Alberti defines the modern *ideal type* of the architect in direct relation to the above transformations of the Renaissance world view.

> I should explain exactly whom I mean by an architect; for it is no carpenter that I would have you compare to the greatest exponents of other disciplines. ... Him I consider the architect, who by sure and wonderful reason and method, knows both how to devise through his own mind and energy, and to realize by construction, whatever can be most beautifully fitted out for the noble needs of man.[4]

Alberti, putting forward the central importance of the power of the human mind and reason for a new definition of the Renaissance architect, is naturally led towards a novel conceptual structure that ascribes meaning to architectural creation.[5] Architecture is divided into *conception* and *construction*. This dualism is projected onto the nature of the object of architectural knowledge itself. As Alberti writes, "the building is a form of body, which like any other consists of lineaments and matter, the one the product of thought, the other of Nature".[6] Consequently, the concept of *lineamentis* exemplifies the meaning of architecture as an ideal conception within the mind of the architect-creator. In this context, the final architectural work, the spatial body of the building, is produced when the builder moulds natural matter according to the *lineamenta*.

Lineamenta: Towards a new interpretation

Many scholars of Alberti have formulated various interpretations of the term, proving its inherent complexity and ambiguity.[7] It seems that according to the most widely held opinion, the concept of *lineamentis* refers to the materialized *design* and *drawing* of the architectural work, namely the represented set of lines and angles that concentrate the basic characteristics of the ground plan, the general geometrical structure and the external form of a building.[8] Caroline van Eck has argued that *lineamenta* designates, apart from the designed drawings, a "*mental activity of planning*".[9] Dalibor Vesely has proposed the phrase "*imaginary or ideal structure of design*" as an interpretation of the term.[10] Françoise Choay has interpreted *lineamenta* as the *form* of construction.[11] Alternatively, she refers to it as the *figure* of the construction.[12]

Following the above, it is argued that the concept of *lineamentis* is concerned with the process of the *pure mental conception* of the architectural work, *before* this process – that is, the *project* – is inscribed and codified into a sensible, designed representation. In other words, in addition to materialized *design*, *lineamenta* seems to have another, more profound and primary meaning. The first proof that can support our thesis is that Alberti uses very obviously another term to denote the designed lines of architectural representations: the term *lineas*.[13] A second proof that strengthens our interpretation is a comment by the humanist in the first book of his treatise. Alberti maintains that all the power and reason of *lineamentis* can be summarized in the guiding method that leads with absolute certainty into the right synthesis of lines and angles which comprehend and complete the functional type, the geometrical shape and the external form of the architectural work.[14] In the above definition, it becomes obvious that *lineamenta* are not identical and do not coincide with the designed lines and angles of the work, but rather ground the normative framework for their justification, the mental scheme that secures their logical validity. A third and last proof in favour of our proposal comes from the following words and phrases of Alberti himself. He writes:

> Nor do lineaments have anything to do with material,[15] … lineamentum is a certain, constant prescription, conceived in the mind.[16] … It is quite possible to project whole forms in the mind without any recourse to the material.[17]

Following the above analysis, it is argued that *lineamenta* have the aim of prescribing and predetermining, *within the architect's mind*, the reason, the type, the purpose, the geometrical space and the final form of the projected architectural work.[18] In relation to the argument so far and bearing in mind the humanistic character of Alberti's world view, a possible new interpretation of the meaning of *lineamenta* could be that they signify *the ideas, the mental concepts and the spatial types* of the project that ascribe an ideal social purpose to an architectural work.

In other words, *lineamenta* concentrate a complex intellectual and mental programme for architectural praxis, schematizing the metaphysical foundations and the logical prerequisites of the architectural endeavour as a whole. Consequently, the process of forming them is placed conceptually and epistemologically at the beginning of architectural creation, acting as an abstract ground for its legitimization. Alberti, placing the treatment of *lineamentis* in the first book of his treatise,[19] recognized their ontological and mathematical importance,[20] their constitutional role for a general meaning of architectural creation.

The social role of architectural representations

The humanistic perspective of Alberti's architectural thought ascribes a new function to architectural representations in the process of architectural creation. Thus,

Alberti, before the investigation of the principles for the correct preparation of the materials for the final construction of the building, maintains that the initial conceptual determinations of architectural thought, as concentrated by its *lineamenta*, ought to be inscribed, described and recorded through a process of *semeiosis* (*perscriptio*).[21]

According to Alberti, the description and inscription of *lineamentis* can be achieved through the drawing and the model. In other words, the design by hand and the construction of models allow for the presentation of many dimensions of the architectural ideas into sensible form. This process has an important function: it provides for a better social, scientific and constructional test and control of architectural ideas.[22] Through architectural representations, the meaning of the architectural conception of *lineamentis*, as created within the architect's mind, is communicated to the public sphere of the city. Codifying the ideal content of the architectural work into an inter-subjective space of reference, architectural representations answer the demand for a collective communication of ideas and social dialogue. In other words, it is argued that for Alberti, the sensible representation of architectural ideas re-inscribes the abstract conception of the architectural work into a framework of concrete historical, cultural and natural laws and conditions, bridging the mental creativity of the individual architect with the collective demands of society as a whole.

In this respect, Alberti consults the architect to represent with the utmost precision, simplicity and objectivity his initial ideas into drawings and models, in order to present accurately and amply the range of his thought for public criticism. Moreover, the sensible representation of abstract ideas contributes decisively to their complete elaboration and their final formulation.[23] In this way, architectural representations condense the original conceptual structures of the "*mental edifice*"[24] into sensible traces and symbolic lines and at the same time refer to the final work under construction, depicting the basic objective order for the organization of built space.[25] Modern architectural representations, as conceived in the founding text of Alberti, create a *social space of mediation* between architectural thought and architectural praxis, a field that bridges with validity and objectivity the epistemological dualism of architecture as conception and construction, unifying *lineamenta* with matter, and connecting the creative subject with the intended object.

The ontological and epistemological primacy of architectural ideas

The origin of architecture as a mathematical and spatial organization of the cosmos can be traced back into a fundamental natural capacity and deeply rooted ability of the human mind itself, which, according to Alberti, always has the tendency to conceive perfect and complete ideas, concepts and forms of buildings, independently of matter, and to project intellectual solutions to building problems.[26]

This creative and spontaneous energy of the human mind can be considered as the source of the building's ideas, which are produced solely by the human mind itself.[27] Man's productive ability to create ideas in the mind can be considered as the ultimate ground for the legitimization, coherence and meaning of architectural creation. This energetic, productive and conscious action of the human mind has a social and ethical character,[28] and, consequently, the intentions and values of the architect-humanist assume a decisive role in the shaping of meaningful spaces for civic life.

In this process, the concept of *lineamentis* expresses this new creative power of the mind, an ability and capacity to conceive ideas and concepts that precede architectural representations, architectural design and architectural praxis. In this perspective, architectural ideas of the human mind have an ontological and epistemological primacy over their representations and act as foundations of architectural creation.

Leon Battista Alberti and Nicholas of Cusa: Conceptual affinities

If the interpretations expounded so far are valid, it could be argued that Leon Battista Alberti's thought presents some analogies with the speculative theories of the Renaissance philosopher Nicholas of Cusa.[29] Cusa, in his work *Idiota de Mente*, conceives the divine and the human mind as sources for the boundary and measure of everything, namely as *measurement* (*mensurare*).[30] The mind has the power to enfold conceptually the exemplars of all external things,[31] namely the archetypes or ideas that constitute their essential form, which, however, do not derive from the senses and are wholly separated from matter.[32] In that way, Cusa claims that the mind's ideas or exemplars precede perceptible things as an original precedes an image and, moreover, that they derive from the simplest, most precise and infinite form or idea, namely, God.[33] Consequently, human mind is structured in direct analogy to the divine mind: according to Cusa, since the latter is the all-encompassing unity of the true nature of things and an exemplar of our minds, the human mind is an all-encompassing unity of the assimilation of things, namely an all-encompassing unity of *concepts*. Following Cusa's thinking, the difference between God's mind and the human mind is that the former produces things while conceiving them, whereas the human mind only *conceptualizes* them.[34]

It is argued that the understanding of the creative power of the human mind, the *power of judgement*, to produce *concepts* of external, sensible things that concentrate their exemplary structures and forms, as propounded by Cusa, reminds of the analogous role of the *lineamenta* in shaping and prefiguring the place, the exact number, the order and the scale of the sensible body and space of a building, as put forward by Alberti.[35] Moreover, the likeness of Cusa's mental conceptions to mathematical notions and numbers presents another similarity with the seemingly mathematical and geometrical dimension of the concept of *lineamentis*, as codified by Alberti.[36]

In other words, the role and the function of *lineamentis* in Alberti's architectural thought presents some affinities with the role and the function of the *concepts* in Cusa's speculative theology and epistemology, if they are both understood as symbolic expressions of the power of judgement that is inherent in the human mind. Alberti and Cusa understand *lineamenta* and *concepts* respectively as the symbolic functions that lead to the creation of a *mental space* of clear ideas, values, harmonious structures, exemplars, paradigms and prototypes that serve as an absolute scale of reference, critique, measurement, organization, assimilation and evaluation of the magnitude, multitude, order, shape, proportions and boundaries of sensible, external and material forms.[37]

Consequently, both Alberti and Cusa recognize the creative dimension of mental, mathematical, geometrical and conceptual structures of the human intellect and their priority for the meaningful elaboration of real, material space and the significant construction of an ordered, structured, coherent and harmonious social and cultural environment.[38] Cusa explicitly refers to the art of building, stating that without the conceptual and creative power of the mind no configuration can be made and no boundaries of things can be marked off.[39] The concept of *lineamentis* expresses the same formative, independent power of the mind, which, according to Cusa, is very different from the surmises of reason that are attached to perceptible material objects. Cusa argues that the original constructive principle of the human mind makes assimilations of forms as *they are in and of themselves*, namely as abstract forms and ideas, without any material reference, using them as exemplars and measures of truth for the respective material and corporeal entities. For example, a geometrical circle conceived within the human mind is immutable and perfect, acting as a measure of truth for a materialized circle in a patterned floor.[40]

The concept of *lineamentis* as expounded by Alberti presents a striking similarity with the above description of the mathematical power of the human intellect to conceive abstract ideas through which the material construction of sensible objects in space can be measured and delineated. In this respect, Alberti's insistence that *lineamenta* are wholly separate from matter acquires its logical explanation and its philosophical justification in the works of Cusa.[41] This conceptual affinity between Alberti and Cusa enhances our interpretation of *lineamentis* as the ontological and epistemological ground that legitimizes the architectural endeavour as a whole and the corresponding priority of certain and necessary intentions and ideas[42] within the architect's mind,[43] over their sensible and designed representations and traces.

Humanistic ideals and *lineamenta*: Alberti and Bruni

The concept of *lineamentis*, as understood by Alberti, is not only characterized by a mathematical, epistemological and geometrical meaning, but, since it is produced by the mind of a creative subject, presents an *ethical, political and social meaning* as well, a meaning that is somehow inherently present within the constitution, the positive intentions, the moral humanism and the optimistic anthropocentrism that accompany this new kind of historical actor, namely the modern architect.[44] Alberti claims that since a new concept of a building is formed in the mind of a humanist citizen, he will "gladly and willingly offer and broadcast his advice for general use, as if compelled to do so by nature."[45] It is argued that this communicative, well-intentioned, inter-subjective and positive social quality of the *lineamenta*, namely the building's ideas and purposes, reveals its ethical and political dimension which enhances the honour and glory of the builder, the city and human society as a whole. The inherent social, political, ethical and altruistic dimension of *lineamentis* is absolutely essential for Alberti's architectural thinking, and provides for its unique humanistic character: it is this dimension that prevents *lineamenta* from becoming mere subjective images or arbitrary visual appearances. On the contrary, the political and ethical quality of the *lineamentum* lends it its objective and general validity, which springs from its constant reference to social, common and collective ideals, values, needs and ends. The humanistic and ethical dimension of *lineamentis* is ascertained by the fact that the use of this term can be traced in the writings of another great humanist of fifteenth-century Italy: Leonardo Bruni. As Michael Baxandall records, Bruni, in his text "*De Interpretatione Recta*", written around 1420, where he considers problems in literary criticism and the theory of translation and interpretation of written texts, compares the painter with the translator and holds the view that the latter should try to recreate all the *lineamenta* of the original author's discourse, as the former restructures the whole idea, form and figure of another painted model.[46]

Bruni clearly uses the words "*mente et animo*" (mind and soul) to refer to this recreative process of the translator, reminding us of the following phrases of Alberti: "[W]e cannot prevent our minds and imagination from projecting some building or other" and "It is quite possible to project whole forms in the mind without any recourse to the material."[47] Alberti uses these expressions to describe the intellectual conception of the *lineamentis* from the creative and recreative capacity of the architect's mind. Those striking similarities reveal that a possible source of Alberti's concept of *lineamentis* is Leonardo Bruni's humanistic, civic and literary thought, which uses original visual metaphors that are applicable both to painting and to literature, understood as different modes of a common public, social and civic discourse.[48]

The ideals of the political citizen and the honour of the public man and the values of citizen participation in the government of the republic, along with a new conception of the life of active political engagement, are all exemplified in Bruni's writings and form an organic part of Florentine Humanism in the fifteenth century. Those humanistic ideals are transferred by Alberti into the sphere of architecture and, especially, in the social, political and ethical concept of *lineamentis*. Hans Baron has written a thorough analysis of the relations between Alberti's thought and the communal ethics of civic humanism.[49] Those ideals are put forward in the Introduction of "*De Re Aedificatoria*", where architecture is conceived as a public endeavour that contributes decisively to the honour and the dignity of society as a whole.[50] The *lineamenta* symbolize and capture those moral values and qualities,[51] concentrating, apart from the form and the geometrical structure of a building, its social, political and ethical character, purpose and destination. Alberti writes:

> Since buildings are set to different uses, it proved necessary to inquire whether the same type of lineaments could be used for several; we therefore distinguished the various types of buildings.[52]

In this remark it is shown that *lineamenta* also define the intended social function and destination of a building, namely its conceptual meaning, character and purpose. According to its linguistic usage in Latin, *lineamentum* also designates the character of the face or the character of the soul, thus having an ethical dimension as well.[53] This ethical dimension of *lineamentis* is used in Cicero's *De Finibus Bonorum et Malorum*, referring to the character of the soul (*animi liniamenta*).[54] It is argued that the social and ethical meaning of *lineamentis*, which prescribe the functional and civic character of a projected building within the human mind, relocates architectural thought, architectural representation and architectural praxis into a broad cultural, philosophical, poetic and ethical discourse.

Contemporary architectural theory and praxis, through an original interpretation of the concept of *lineamentis*, could possibly restore the humanistic content of architectural creation and transform architectural representations from mere digital or representational images to *image-worlds*, according to Ernst Cassirer's meaning of this term,[55] thus symbolizing general cultural and philosophical discourses. This enlarged and more profound notion of architectural creation, based on general ideals and concepts, could enhance the interdisciplinary dialogue between intellectual inquiry in the humanities and design creativity, leading to the thoughtful realization of meaningful spaces that will improve the ethical, symbolic and social quality of everyday human life.

Notes

1. See the critical edition of the original Latin text: L.B. Alberti, *L'Architettura [De Re Aedificatoria]*, Testo latino e traduzione a cura di Giovanni Orlandi, Introduzione e note di Paolo Portoghesi, Tomo 1, 2, Milano: Edizioni Il Polifilo, 1966.
2. In this respect see especially the research programme of *Dalibor Vesely*, as expounded in the book D. Vesely, *Architecture in the Age of Divided Representation, The Question of Creativity in the Shadow of Production*, Cambridge, MA, and London: The MIT Press, 2004, pp. 3–8.
3. See J. Burckhardt, *The Civilization of the Renaissance in Italy, An Essay*, London: Phaidon Press Ltd, 1965, pp. 1–103; E. Cassirer, P.O. Kristeller and J.H. Randall Jr., (eds), *The Renaissance Philosophy of Man*, Chicago, IL, and London: University of Chicago Press, 1948; E. Cassirer, *Le Problème de la Connaissance dans la Philosophie et la Science des Temps Modernes*, I, Traduit de l'allemand par René Fréreux, Préface par Massimo Ferrari, traduite de l'italien par Thierry Loisel, Paris: Les Éditions du Cerf, 2004, pp. 69–73; E. Cassirer, *The Individual and the Cosmos*

in Renaissance Philosophy, translated with an introduction by Mario Domandi, Mineola, NY: Dover Publications Inc., 2000, pp. 1–6, 24–36.
4 See L.B. Alberti, *On the Art of Building in Ten Books*, translated by Joseph Rykwert, Neil Leach and Robert Tavernor, Cambridge, MA, and London: The MIT Press, 1988, p. 3. On the original Latin text see L.B. Alberti, *L'Architettura [De Re Aedificatoria]*, p. 7.
5 L.B. Alberti, *L'Architettura [De Re Aedificatoria]*, pp. 7–11.
6 L.B. Alberti, *On the Art of Building in Ten Books*, p. 5. On the original Latin text see L.B. Alberti, *L'Architettura [De Re Aedificatoria]*, p. 15.
7 For a discussion of the ambiguous nature of the concept of *lineamentis* and its resemblance to the Mannerist's notion of *disegno interno* see D. Vesely, *Architecture in the Age of Divided Representation, The Question of Creativity in the Shadow of Production*, pp. 133–38. See also L.B. Alberti, *L'Art d'Édifier*, Texte traduit du latin, présenté et annoté par Pierre Caye et Françoise Choay, Ouvrage traduit et publié avec le concours du Centre National du livre, Paris: Éditions du Seuil, 2004, p. 55, n. 1.
8 See L.B. Alberti, *On the Art of Building in Ten Books*, pp. 422–23, S. Lang, "De Lineamentis: L.B.Alberti's Use of a Technical Term", *Journal of the Warburg and Courtauld Institutes*, 1965: XXVIII, pp. 331–35, A. Pérez-Gómez and L. Pelletier, *Architectural Representation and the Perspective Hinge*, Cambridge, Massachusetts, London, England: The MIT Press, 1997, p. 9, E. Panofsky, *Idea, Contribution à l'Histoire du Concept de l'Ancienne Théorie de l'Art*, Traduit de l'Allemand par H. Joly, Préface de J. Molino, Paris: Editions Gallimard, 1989, pp. 214–215, n. 125, L.B. Alberti, *L'Architettura [De Re Aedificatoria]*, p. 18, E. Kaufmann, *Architecture in the Age of Reason, Baroque and Post-Baroque in England, Italy, and France,* New York: Dover Publications, Inc., 1968, p. 93.
9 C. van Eck, "The Structure of *De Re Aedificatoria* Reconsidered", *Journal of the Society of Architectural Historians,* 1998: 57, 3, n. 15.
10 D. Vesely, *Architecture in the Age of Divided Representation, The Question of Creativity in the Shadow of Production,* p. 139.
11 F. Choay, La Règle et le Modèle, sur la Théorie de l'Architecture et de l'Urbanisme, Paris: Éditions du Seuil, 1980, pp. 92–93.
12 F. Choay, *The Rule and the Model, On the Theory of Architecture and Urbanism*, edited by Denise Bratton, Cambridge, MA, and London: The MIT Press, 1997, p. 69, 72, 327 n. 15.
13 L.B. Alberti, *L'Architettura [De Re Aedificatoria]*, pp. 15–21.
14 The attribution is mine, from the original Latin. See L.B. Alberti, *L'Architettura [De Re Aedificatoria]*, p. 19. I interpret the term *facies* in a threefold way, namely as type, shape and external form. For the basis of this interpretation see S. Koumanoudis, *Latin-Greek Dictionary*, Athens: Grigori Publications, 2002, p. 296.
15 See L.B. Alberti, *On the Art of Building in Ten Books*, op. cit., p. 7. On the original Latin text see L.B. Alberti, *L'Architettura [De Re Aedificatoria]*, pp. 19–21.
16 The translation is mine, from the Latin text in L.B. Alberti, *L'Architettura [De Re Aedificatoria]*, p. 21.
17 See L.B. Alberti, *On the Art of Building in Ten Books*, op. cit., p. 7. On the original Latin text see L.B. Alberti, *L'Architettura [De Re Aedificatoria]*, p. 21.
18 L.B. Alberti, *L'Architettura [De Re Aedificatoria]*, p. 19.
19 *Ibid.*, pp. 21–93.
20 V.P. Zubov, "La Théorie Architecturale d'Alberti [Arhitekturnaâ teoriâ Al'berti]", in *Albertiana, Volume III*, Firenze, Italia: Leo S. Olschki Editore, 2000, p. 21.
21 L.B. Alberti, *L'Architettura [De Re Aedificatoria]*, p. 53.
22 *Ibid.*, pp. 97–99.
23 *Ibid.*, pp. 97–109.
24 M. Frascari, "The Tell-the-Tale Detail", in *Theorizing a New Agenda for Architecture, An Anthology of Architectural Theory 1965–1995*, K. Nesbitt (ed.), New York: Princeton Architectural Press, 1996, p. 503.
25 L.B. Alberti, *L'Architettura [De Re Aedificatoria]*, p. 173
26 On the original Latin text see L.B. Alberti, *L'Architettura [De Re Aedificatoria]*, p. 11.

The human mind and design creativity

27 In this respect, see the Latin expressions: "*mentem cogitationemque*" and "*ab ingenio produceretur*" in L.B. Alberti, *L'Architettura [De Re Aedificatoria]*, p. 15.
28 See L.B. Alberti, *L'Architettura [De Re Aedificatoria]*, p. 11.
29 For a general biographical account of Cusa's life and works see J. Hopkins, "Nicholas of Cusa", in *Dictionary of the Middle Ages*, Vol. 9, J.R. Strayer (ed.), New York: Charles Scribner's Sons, 1987, pp. 122–25.
30 Nicholas of Cusa, "Idiota de Mente" [The Layman on Mind], in *Nicholas of Cusa, on Wisdom and Knowledge*, J. Hopkins (ed.), Minneapolis, MN: The Arthur J. Banning Press, 1996, pp. 535–36.
31 Nicholas of Cusa, "Idiota de Mente" [The Layman on Mind], p. 536.
32 *Ibid.*, pp. 538–40.
33 *Ibid.*, pp. 540–42.
34 *Ibid.*, pp. 543, 556.
35 Nicholas of Cusa, "Idiota de Mente" [The Layman on Mind], pp. 544–46, 551–55 and L.B. Alberti, *On the Art of Building in Ten Books*, p. 7.
36 Nicholas of Cusa, "Idiota de Mente" [The Layman on Mind], *op. cit.*, pp. 554, 565–69 and L.B. Alberti, *On Painting and on Sculpture*, the Latin texts of *De Pictura* and *De Statua*, edited with translations, introduction and notes by Cecil Grayson, London: Phaidon Press Limited, 1972, pp. 36–37.
37 See Nicholas of Cusa, "Idiota de Mente" [The Layman on Mind], pp. 570–71; E. Cassirer, *The Individual and the Cosmos in Renaissance Philosophy*, pp. 7–51: E. Cassirer, *Le Problème de la Connaissance dans la Philosophie et la Science des Temps Modernes*, I, pp. 29–46; G.F. Vescovini, "Nicholas of Cusa, Alberti and the Architectonics of the Mind", in *Nexus II, Architecture and Mathematics*, Williams, K. (ed.), Fucecchio: Edizioni dell'Erba, 1998, pp. 159–71; K. Harries, "Cusanus and the Platonic Idea", *New Scholasticism*, 1963: 37, pp. 190–91, 200–1, K. Harries, *Infinity and Perspective*, Cambridge, MA and London: The MIT Press, 2001, pp. 187–90.
38 K. Harries, "Cusanus and the Platonic Idea", pp. 199–201 and K. Harries, *Infinity and Perspective*, pp. 190–99.
39 Nicholas of Cusa, "Idiota de Mente" [The Layman on Mind], p. 557.
40 *Ibid.*, p. 558.
41 For a discussion of the relation between Alberti and Cusa see K. Harries, *Infinity and Perspective*, pp. 64–70, 340 n. 6 and D. Vesely, *Architecture in the Age of Divided Representation, The Question of Creativity in the Shadow of Production*, p. 156.
42 Nicholas of Cusa, "Idiota de Mente" [The Layman on Mind], *op. cit.*, p. 559 and K. Harries, "Cusanus and the Platonic Idea", pp. 191–92.
43 For this notion of ideas as conceptual intuitions within the mind of the artist or the architect see E. Panofsky, *Idea, Contribution à l'Histoire du Concept de l'Ancienne Théorie de l'Art*, pp. 22–23 and D. Vesely, *Architecture in the Age of Divided Representation, The Question of Creativity in the Shadow of Production*, pp. 161–63.
44 C. Grayson, "The Humanism of Alberti", *Italian Studies*, 1957: XII, pp. 37–56 and K. Harries, *Infinity and Perspective*, pp. 185–190.
45 See L.B. Alberti, *On the Art of Building in Ten Books*, p. 4. For the original Latin text see L.B. Alberti, *L'Architettura [De Re Aedificatoria]*, p. 11.
46 M. Baxandall, *Giotto and the Orators, Humanist Observers of Painting in Italy and the Discovery of Pictorial Composition, 1350–1450*, London and Oxford: The Clarendon Press, 1971, pp. 22–25.
47 See L.B. Alberti, *On the Art of Building in Ten Books*, pp. 4, 7. On the original Latin text see L.B. Alberti, *L'Architettura [De Re Aedificatoria]*, pp. 11, 21.
48 M. Baxandall, *Giotto and the Orators, Humanist Observers of Painting in Italy and the Discovery of Pictorial Composition, 1350–1450*, p. 26.
49 H. Baron, *In Search of Florentine Civic Humanism: Essays on the Transition from Medieval to Modern Thought*, Vol. I, Princeton, NJ: Princeton University Press, 1988, pp. 258–78.
50 L.B. Alberti, *L'Art d'Édifier*, p. 50.
51 For a relation between Alberti's thinking and moral thought, especially Cicero's, see J. Onians, "Alberti and ΦΙΛΑΡΕΤΗ, A Study in their Sources", *Journal of the Warburg and Courtauld Institutes*, 1971: XXXIV, pp. 96–114.

52 See L.B. Alberti, *On the Art of Building in Ten Books*, *op. cit.*, p. 5. On the original Latin text see L.B. Alberti, *L'Architettura [De Re Aedificatoria]*, *op. cit.*, p. 15.
53 See S. Koumanoudis, *Latin-Greek Dictionary*, p. 484.
54 Cicero, *De Finibus Bonorum et Malorum*, Cambridge, MA and London: Harvard University Press, 1914, pp. 294–95.
55 For the notion of the *image-world* in relation to the formative power of the human spirit see E. Cassirer, *The Philosophy of Symbolic Forms, Volume One: Language*, New Haven and London: Yale University Press, 1955, p. 78.

Chapter 13

Renaissance visual thinking

Architectural representation as a medium to contemplate "real appearance"

Federica Goffi-Hamilton

Introduction

During the Renaissance architectural drawing was understood as a medium to contemplate "real appearance". Real appearance was not just a representation of *likeness* but also an epiphany of *presence*. An analysis of the 1571 hand-drawn ichnography of St. Peter's Basilica in the Vatican, by Tiberio Alfarano, provides clues about the weaving of ideas into drawing.

Alfarano's work is porous to the cultural context in which it was produced. The scholar of the basilica's history, theologian and connoisseur of architecture wove into the drawing a complex body of religious, political, architectural and cultural elements. The 1571 ichnography is – in Carlo Ginzburg's terms – a "singularity", presenting a series of anomalies when compared with the surviving body of renovation drawings (1506–1626).[1] A key anomaly is the fact that Alfarano was not, strictly speaking, an architect and this is not a design drawing per se. This perhaps explains why this representation has been largely overlooked by recent scholarship.

The drawing exhibits advanced representation techniques, which escape more traditional architectural drawing methods. One has to wonder as to why Alfarano created a unique hybrid representation combining *hand drawing* with *decoupage* and used rendering techniques of *icon painting*. The beneficiary clerk deviates from rational types of architectural representation, used at the time, to define a completely new and non-rational kind of representation. This hybrid drawing – merging techniques used in the making of sacred texts with traditional architectural representation

– requires an appropriate reading method to reveal the sacred meanings embedded. The method entailed a relationship between word and image that was typical of a sacred text.

Even though acknowledged as an essential "document" by scholars, because of the information about the location and translation of important relics and altars from old to new St. Peter's, the drawing has not been examined in terms of silent meanings expressed visually through the use of iconographic language typical of icon paintings, which go well beyond the pragmatic level. In this regard its significance is still under-evaluated, and its potential influence on architectural thinking yet to be uncovered.

Alfarano provides a narrative of the Renaissance transformation that goes beyond witnessing geometrical exactness to describe physical form, and powerfully enters into a much deeper and profound realm, that of the theological significance of the transformation. He explains the relationship between old and new St. Peter's through a sophisticated spatial/temporal relationship between the measured/scaled drawing of superimposed plans, framed by a series of icons critically positioned in relationship to the architectural body of the temple.

Such complex narrative is developed making use of a language that was well known to Alfarano, that of sacred texts, where the interdependence between word and image reveals a complex relationship between drawing and building. The drawing becomes the effigy of the temple, i.e. a twinned body, synonymous to the

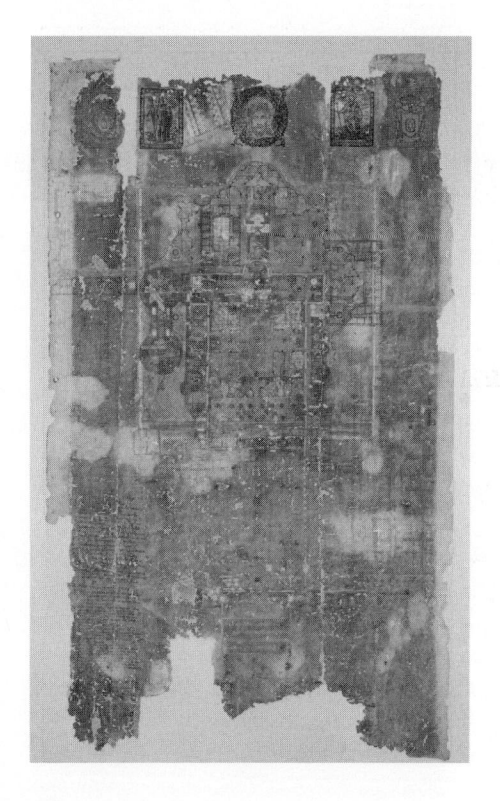

Figure 13.1
Tiberio Alfarano. Hand drawing, 1571 (dimensions 1172 mm × 666 mm). © Courtesy of the Archivio della Fabbrica di San Pietro.

martyred body of the Church, offering itself as a *substitute* allowing the experiencing of real presence.[2] This is possible because of the author's understanding of the relationship between the humanities and visual form, derived from the understanding of the making and reading of sacred texts.

Penetrating into the drawing, and inquiring into its metaphorical significance, allows opening up deeper levels of interpretation beyond the pragmatic. This in turn provides preliminary answers as to the time dimension of this drawing, and the processes that allow Alfarano to embed his complex interpretation and narration of transformations – happening and still to take place – in Renaissance St. Peter's.

Alfarano claims to represent the "real appearance" (*forma sacrosanctae*), i.e. an iconic portrait, embodying an atemporal essence beyond one-time likeness. The 1571 hand drawing is conceived as an "iconic portrait," i.e. a metaphoric gate providing access into a spiritual realm, demonstrating the essence of that which is symbolized, beyond a mere portrayal of likeness. The drawing leads the contemplating viewer beyond the gate of the visible, into a transfixed realm, to reveal a veiled significance.[3]

While reflecting counteracting views on the future of the basilica, the drawing provides Alfarano's influential viewpoint regarding the ongoing debate on the most appropriate plan, i.e. the Greek or the Latin cross. Such well-known controversy has been amply discussed by scholarship.[4] A critical analysis of the clues hidden in the drawing indicates the presence and necessity of a hybrid body merging two cross types.

A critical analysis of Renaissance visual thinking, might inform present theory and practice of architecture. In today's understanding, architectural drawing projects an image of likeness. As such, representation renounces the dialogue with the humanities and becomes a narcissus, i.e. a self-reflection of the visual world, projecting a still image rather than an *image of becoming*. The "dominance of image" as the only legitimate way to generate design ideas should be challenged by undermining the notion that architectural drawing is a portrayal of likeness, restoring its full potential to represent "real appearance," i.e. an iconic representation of presence.

Disassembly and *rebirth* of St. Peter's Basilica

During the sixteenth century and the beginning of the seventeenth century, the so-called "Old" St. Peter's incorporated a series of transformations guided by the hands of several architects and popes. In a period of 120 years – starting in 1506 under the pontificate of Julius II (1503–1513) and coming to completion in 1526 with the addition of a new facade under Paul V (1605–1621) – the body of the Vatican temple was entirely renovated. The reconsecration was held on 18 November 1626.

St. Peter's regeneration involved a gradual disassembly, and its rebirth *as* "new" St. Peter's. Alfarano's drawing (Figure 13.1), executed in a sort of *middle point* within the timeline of historical and architectural changes, is of the greatest significance in order to understand the meaning of such changes, and connect the *before* with the *after*, revealing the *continuum* formed by old and new basilicas.

The disassembly of the building spoil by spoil, relic after relic, happened simultaneously to the new construction. The two bodies – new and old – joining, intertwining and overlapping, were never truly separate nor could they be said to be so today. Not even the so-called "dividing wall" built in 1538, erected between the new construction site (surrounding the *confession*) and the eastern portion of the old basilica, truly separated them.[5] On the western side, the tomb of the Apostle Peter, the Constantinian *pergula* and the old basilica's apse survived. These were incorporated within – and protected by – Donato Bramante's *tegurium*.[6] Portions of the transept were still erected, meanwhile, the original floors lay concealed underneath a layer of construction site materials.

Alfarano refers to the dividing wall as the "newer wall", but also as "the great wall intersecting new and old".[7] Rather than as an element of separation, the wall was interpreted as a *new* intersecting element, assuring continuity and contiguity. It is a shared *threshold* joining two bodies, suturing the surviving structures to the new, thereby providing stability.

Throughout renovation, maintaining the identity of the temple was essential. The restoration (*instauratio*) was meant to be a conservation, i.e. the making was a remaking.[8] In the manuscript's preface Alfarano explains that, by understanding the new temple's design, his soul was converted to believe that the old temple did not perish with the erection of the new.[9] He explains that the old main nave and transept are clear of new foundations, which are instead located in the secondary aisles outside the original footprint. This allowed for the conservation of the burials placed within the footprint of the Latin cross.

The title of Alfarano's manuscript, "The Oldest and Newest Basilica" implies the presence and joining of two bodies, whose merging is demonstrated through a "single folio", i.e. the *one and only* drawing hand drafted. Alfarano draws two plans, the oldest Latin cross and the newest central plan by Michelangelo, which together form a hybrid. Archeological excavations conducted during the 1940s and 1950s around the "*confession*" brought to light vestiges of the Old St. Peter's, such as truncated columns and walls, including portions of the "*newer wall*".[10] The old basilica did not cease to exist with the erection of the new temple, but rather it was *incorporated* within it. Alfarano makes specific reference to the fact that founding the *new* pilasters within the side aisles allowed the *old* vestiges of the main nave and transept to remain intact. This was a necessity dictated by Julius II, aiming to preserve the holy sepultures underneath the basilica, forming the living cross and revealing the presence of the real body of the Church of Christ.

The hand drawing: Demonstration *in the making*

Tiberio Alfarano's hand drawing with its superimposition of physical and metaphorical layers is a demonstration *in the making* of the presence of two bodies joined together.¹¹ Pope Julius II approved Bramante's "*experiment*" upon strict condition that the old cross be conserved.¹² Alfarano witnesses and measures on-site the old basilica, of which the eastern "half" survives beyond the "newer wall". The drawing is a correct demonstration of the location of the old apse, in relationship to the new piers.¹³ At that time Bramante's *tegurium* and the west portion of Constantine's apse form a hybrid structure protecting the main altar.¹⁴

Alfarano's ability to layer meanings through drawing witnesses his skillfulness, inventiveness and his imaginative understanding of architectural representation. It could be argued that Alfarano's drawing (1569–1571–1576 …) was valued because of his ability to fuse together *both* the architectural and the theological intentions.¹⁵ In his double role of member of the clergy and connoisseur architect, he held a privileged position and a rare viewpoint. He conveys not just a physical survey of new and old elements, but also their metaphorical relationships.

Alfarano literally cut out Étienne Dupérac's 1569 print of Michelangelo's plan and physically glued it onto his hand-drawn plan of the old basilica. This was executed on paper mounted on a wood board, thereby demonstrating the combined existence of both plans. Essentially a decoupage of print clippings of various provenance, Alfarano's drawing could be interpreted as a complement to the incomplete story narrated by Dupérac's print; it records the essence of the transformation shortly after Michelangelo's death, demonstrating not just the *newest* added member (the central plan) but also its relationship with the old one.

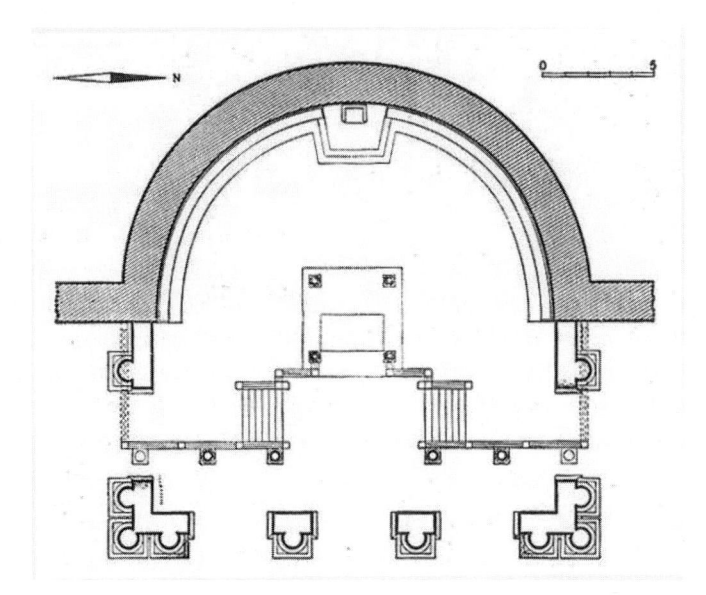

Figure 13.2
Plan reconstruction of Bramante's *tegurium*. It is possible to observe the "hybrid" body formed by portions of the surviving wall of the Constantinian apse and Bramante's "new" addition, framing within it six original columns of the Constantinian *pergula*. This drawing portrays the situation around the area of the *confessio* around Alfarano's time

Through this process Alfarano in effect added his own commentary on to Dupérac's fragmentary history.

Significance of the archeological layers of the drawing

Alfarano's drawing is executed on several sheets of paper of different dimensions, quality and consistency. On this support Alfarano first drew in graphite the Constantinian basilica, in its original integrity. The inscription in the lower-left corner states that Michelangelo's new temple's *ichnographia* is "added" above that of the old.[16] Alfarano glues on the Constantinian plan a cut-out of Dupérac's 1569 print, aligning Michelangelo's central plan to the hand drawing by means of a painter's grid. Subsequently, he redraws on the print the portions of the old basilica hidden by the overlay, highlighting its presence by means of gold paint.[17]

The iconographic presence of Old St. Peter's alludes to the conservation of the main nave and transept's footprint, which had to be preserved. The gold paint alludes to a duality of corpora and spirit – i.e. geometrical likeness and allegorical presence of the real cross. Gold, a material that signifies everlastingness, is used to

Figure 13.3
1532–1536 View from the north of the old transept toward the altar by Martin Van Heemskerck, demonstrating the survival of the screen columns and their architrave (1498–1574), Stockholm, Nationalmuseum, Collection Anckarvärd, n. 637. Pen and ink with white wash on paper

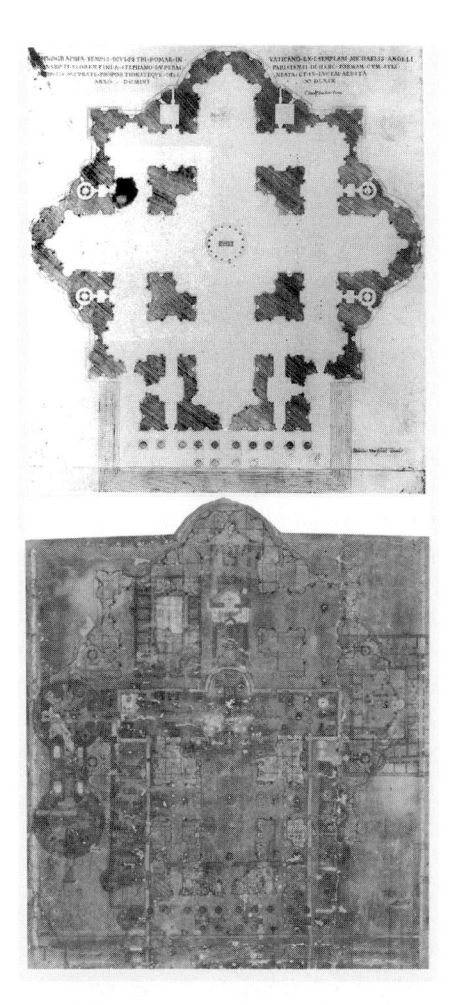

Figure 13.4
Etienne Duperc's 1569 print reproducing Michelangelo's central plan for St. Peter's. Below: A fragment of Tiberio Alfarano's plan (1571) corresponding to the spoil of Dupérac's print collaged onto the drawing. Photoshop alteration of the original by author

allude to both the literal preservation of the original footprint and the presence of the Church as a spiritual entity.

The basilica was, in fact, from the beginning, a multifunctional space combining memorial, burial and liturgical practices.[18] The main nave had been used for centuries as a cemetery. This element, together with Saint Peter's burial, are essential in maintaining material and spiritual continuity. Alfarano's representation reveals the conservation of "totum pavimentum". New St. Peter's is defined by its relationship with the old. A careful dialog between the two plans determines the position of the new piers. The piers in fact should not interfere with the original main nave and transept.

The four piers are placed outside this sacred ground. In 1506 Bramante positioned the eastern piers right under the arms of the crossing – between the old transept and the main nave – indicated in his drawing (U 20A). The north–south distance between them is equal to the width of the old main nave. The challenge for the vaulting of St. Peter's dome is set. The necessity to *remember* the presence of the sacred crossing foretells the pier's position. When the width of the main nave is placed in the east–west direction, the position of the western piers is also foretold. As a result, the tomb of the apostle is not placed in the geometric center of the four piers' crossing.

The old side aisles are interpreted as a peripheral portion of the body, whose vestiges could be retouched. The two eastern piers fall within this area. The western piers instead fall outside Old St. Peter's body, above the shoulders of the cross and outside the original footprint. Saint Peter's tomb is the remainder of the spiritual and material continuity between old and new. Saint Peter's body is the *first stone* on which Jesus lays the foundation of his Church, in both a spiritual and material understanding. In 1571 the new and old basilicas were still joined by the "*newest wall*". When Alfarano drafted his plan only the western half of the Constantinian basilica was unbuilt, surviving as a footprint underneath the new.[19] In anticipation of the "unbuilding" of the eastern portion, Alfarano proposed a design for the addition of an eastern arm, suggesting the necessity to extend the foot of the square cross.[20]

He favored – just like Carlo Borromeo (1538-1584) in his 1577's *Instructiones* – the Latin cross. Alfarano argued that the original footprint should be contained entirely within the new temple. "Real appearance" is revealed in the merging of Latin and Greek crosses – a "hybrid plan" incorporating the original cross and new members. The terminology of "real appearance" (*forma sacrosanctae*), which Alfarano used, is significant. His plan does not portray what was present at the site in 1571. The actual temple's plan would be more like Leonardo Bufalini's map, dating from 20 years earlier.

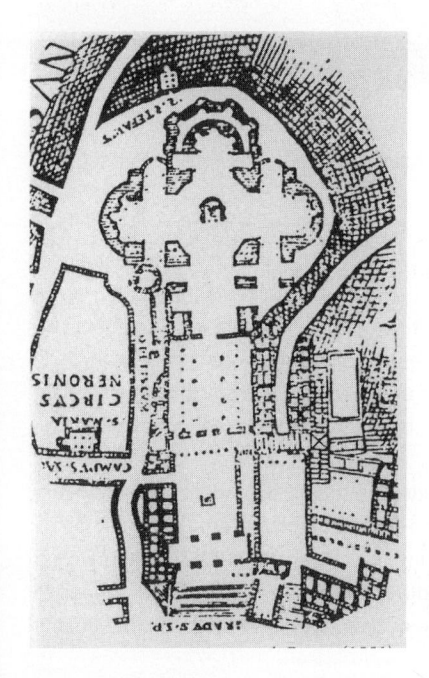

Figure 13.5
Detail of Leonardo Bufalini's map of Rome showing the plan of St. Peter's Basilica in 1551

The Latin word "*forma*", whilst translated as "appearance" is used here to indicate *presence*. The word is charged with significance, which Hans Belting explains relates to something beyond mere likeness – to embody essence.[21] Robert Estienne (1503–1559), in his *Thesaurus linguae Latinae*, gives the Greek correspondent of the Latin form as *eidos, idea, tuttos*. According to Plato in *Phaedo*, *eidos* is the immutable genuine nature of a thing, which gives identity, providing continuity despite the changes that invariably happen to its physical likeness.[22] Leon Battista Alberti (1404–1472), in his treatise "*De Re Aedificatoria*" (1485), makes use of the word "*forma*" in relationship to drawing.

> It is the function and duty of lineaments, then, to prescribe an appropriate place, exact numbers, a proper scale, and a graceful order for whole buildings and for each of their constituent parts, so that the whole form and appearance of the building may depend on lineaments alone.[23]

Alberti states that both the *form* and *appearance* of the edifice rest on the lineaments. Drawing has the potential to reveal not just likeness but presence, not just body but soul in terms of similitude through representation. Hence, Alfarano's drawing should not be interpreted simply in terms of likeness, but also in terms of essence. Moreover, the work should not be treated as a final drawing, or a literal one providing (in modern terminology) a photographic likeness of the represented object. Instead it reveals a program of intentions revealed *in time* through making. The outlined drawing – a "wire diagram" of the thing signified – seems to emphasize the importance given to capturing appearance; combining the figure of a thing with its essence rather than focusing on an unstable one-time likeness.[24]

A careful comparison of Alfarano's hand drawing (1571) with Bufalini's plan (1551) allows to further support the principle that Alfarano's interest lay in representing something other than what we could call *ante litteram* – an instantaneous *still shot*. Bufalini's plan shows the result of the transformations as a literal survey demonstrating an actual layout, i.e. the coexistence of two *unmerged* fragments joined by the "*dividing wall*"; the surviving eastern arm and the new centralized western portion.

Alfarano's hand drawing, on the other hand, is a representation of *process*, showing the *merging* of two plans. He did not just draw that which exists above ground level in 1571 – the physical result of the addition of a new central plan above the old basilica – but rather drew both "footprints" as complete plans and juxtaposed them through metaphoric transparency, demonstrating the merging of the two below ground. This layered juxtaposition is of significance. It appears as if the two plans "blend" within the same space–time frame, due to an intended virtual transparency. This allowed for different temporal layers to be simultaneously perceived. The drawing focuses on the *before* and *after* rather than the *now*; being in essence a representation of becoming.

Alfarano does not draw the result of the architectural "cut" and "paste". He represents the essence of the hybrid body; the result of a transformation where two bodies merge, ultimately emphasizing continuity within change. Alfarano's drawing provides a reading of the renovation as a continuum *work in progress*, which qualifies the building as an unfinished opera unfolding over time and reflected in the making of the drawing as a palimpsest continuously updated.

The gold paint of the Constantinian basilica's walls, layered above Michelangelo's azure walls poché, makes it appear superimposed above Michelangelo's plan. The gold paint – rendered with a procedure similar to that used for illuminated miniature books – is the visible sign of the sempiternity of the old temple, indicating the presence of the "mystical body", i.e. the "spirit" of the Church. The Armenian bole undercoating revealed on the edges of the gold paint, marks the walls' outlines and signifies the presence of the "material body" of the Church in its bloody sacrifice.

In the 1590 printed version of the hand drawing the "*newest wall*" disappears.[25] The old walls on the other hand, which are rendered with solid black ink, come to the foreground. Outlining only the western portion of the new basilica's walls, which fall outside the old footprint, the beneficiary clerk emphasizes the gesture of *circumscribing* the old temple, thereby signifying its conservation within the new.[26]

The presence of Veronica's veil in the top portion of the drawing provides a clue to the theological significance embedded in the drawing. Alfarano alludes to the intention to represent the "*vera icona*", or true effigy, of the temple. The drawing acquires the status of iconic representation. The sacred cloth carried the "imprint" of Christ's face, transferred onto the veil by the blood oozing from his face. Veronica, a woman moved by the sufferance of Christ, offered him the cloth while he was carrying the cross on his way to Golgotha. The process of construction of the Veronica icon in this drawing is the result of the gluing of different print clippings from various provenances. The central piece is a "*bulino*" engraving, representing the traditional cloth, with the head of Christ crowned with thorns. This element is surrounded by a xylographic print clipping of an evergreen garland, framing the veil.[27] Understanding how icons work is a necessary step in reading the drawing. Veronica's veil provides access to an invisible world beyond. Through contemplation *ad faciem* this iconic drawing can be read in its temporal and spatial depths to reveal the qualities and virtues of the drawing. The veil is the instrument which allows revealing "real appearance"; an inviolable atemporal portraiture, capturing the essence of the thing signified beyond a one-time likeness.

Alfarano's decision to represent "real appearance" through *one* drawing is meaningful. He purposefully chose the ichnography as the agency of this revealing. This kind of representation, like the etymology of the word suggests, has iconic qualities which go beyond likeness, implying a relationship of similitude between signified and signifier resolved through appropriate representational strategies.[28] From the Greek *iknos* (track or footprint) and *graphia* (writing), the ichnography becomes a "track drawing" which makes the passage of time visible. The ichnography is thus the chosen instrument to provide memory traces on the "drawing site", which acts like a veil bearing the traces of the building's presence within time.

Alfarano's use of the word *"perstrinxi"* – meaning to "inscribe", in reference to drawing – reminds one of the idea of "wounding" the surface of the paper.

Mary Carruthers explains how writing involved, from the beginning, a kind of physical labor. Sharp instruments were used to wound animals' skins to produce marks. She further argues that during the late Middle Ages the body of Christ is likened to "a written parchment".[29] Alfarano represents, in a non-literal way, the result of a transformation where two bodies – Greek and Latin crosses – coexist and intertwine, emphasizing *continuity within change*, thereby revealing the presence and necessity of a composite plan or "hybrid body".[30]

The iconographic fragments selected and reassembled into a new whole reflect the double nature of Christ as both *human* and *divine*. The image of the savior crowned with thorns, representing the suffering Christ, can be interpreted as an allegory, alluding to the presence of the old basilica; while the evergreen garland surrounding the Holy Cloth is the reminder of the victory over mortality of the resurrected Christ. This could also allude to the new central plan circumscribing the old.

These iconographic elements reveal a double symbolism, indicating both the human and divine nature of Christ, reflected by the two plans. According to Alfarano's ichnography, only the merged presence of the two plans describes the "real appearance". The placement of this icon within the general composition is the hinge to our understanding of the central plan as a symbol of restoration. Alfarano makes reference in the manuscript to the renewal as a "second restoration of the Temple of God".[31] The theological meaning embedded in the overlaid central plan could be argued to be a representation of the "restored" Church of God.

Conclusion

Alfarano's drawing escapes chrono-historical classifications, demonstrating an atemporal representation of substance. His understanding of architectural drawing could be likened to a sacred text where symbolic meanings are revealed through appropriate visual language; by implication, these deeply embedded analogies serve as a powerful discourse witnessing the continuity between word and image, between the humanities and visual form.

The adjective "sacrosanct", used as an attribute to "appearance" in reference to the ichnography, refers not only to its being the most holy but also to its inviolability. The description under the drawing reads: "This is the intact ichnography of the very old Temple of St. Peter Apostle and Prince". The Latin adjective "*integra*" refers to the wholeness of the temple, preserved in metaphoric integrity.

Old St. Peter's has an ineludible presence, witnessed in Alfarano's drawing by using gold paint in the walls' renderings. The pavement of the old basilica was maintained throughout the renewal, and restored in 1574.[32] This has been interpreted as the intention to maintain the eastern portion of the old basilica. It is to be observed though that, after Michelangelo's death in 1564, Antonio Da Sangallo raised the level of the new pavement on the western side beyond the "*newest wall*". The restoration could be interpreted as the intention to maintain – albeit by concealed means – the iconic vestiges, i.e. the "footprint" revealing the presence of the living body of the "Church of Christ". The pavement was treated as a relic that was hidden but not erased. Alfarano's drawing becomes the instrument that makes visible the invisible presence of Constantine's basilica, underneath and inside the new.

Alfarano kept working at his drawing as a palimpsest, as a continuous work in progress until 1576. The plan demonstrates the location of newly translated altars, relics and other elements of significance in both their original locations and the new ones simultaneously.[33] The drawing exists in a "multi-temporal" dimension, looking in the direction of both "time past" and "time future". He never erased the old location of a translated element, but rather kept the memory of it. By so doing Alfarano underlines the relationship between old and new, and the continuity between them, by emphasizing the basilica's perdurance.

Alfarano's drawing has been defined as a "rough draft", i.e. a preliminary drawing for a possible "final drawing". It has even been interpreted as a "preparation drawing" for a future print.[34] The drawing should instead be viewed as an *original*, which is always *in the making* and carries all the marks of its transformation like the basilica's own body. The complete story could only be told by looking simultaneously in two directions in time, providing a representation of process. In today's practice, drawing should enhance our ability to read beyond its immediate appearance by revealing the essence of the portrayed object that transcends a single spatio-temporal condition.

Architectural drawing should therefore be interpreted as a "palimpsest"; an unfinished text carrying the marks of its making and remaking that unfold as a continuous process of layering different strata of meaning. This includes – but is not limited to – geometrical exactness. Alfarano's work reveals the imagination of conservation; new St. Peter's is not a "new" building erected from scratch, but rather a continuation of the "old".

Notes

1   Carlo Ginzburg, "Microhistory: Two or Three Things that I Know about it", *Critical Inquiry* 1993: 20–21 (Autumn), pp. 10–35.
2   Federica Goffi, "Architecture's Twinned Body: Building and Drawing", in Marco Frascari, Jonathan Hale and Bradley Starkey (eds.), *From Models to Drawings, Imagination and Representation in Architecture*, New York: Routledge, 2007, pp. 88–98.
3   Pavel Florensky describes the operating mode of icons as a kind of transference which opens a window towards "ontological self-identity". Florensky Pavel, *Iconostasis*, New York: St. Vladimir's Seminary Press, 1996, pp. 65, 71–73.
4   Arnaldo Bruschi, Christoph Luitpold Frommel, Franz Wolff Metternich, Christof Thoenes, *San Pietro che non c'è, da Bramante a Sangallo il Giovane*, Milano: Electa, 1996.
5   The locution of "new" and "old" St. Peter's came in use after 1538, with the erection of the wall. Ennio Francia, *1506-1606: Storia della Costruzione del Nuovo San Pietro*, Roma: De Luca Editore, 1977, p. 51.
6   The *tegurium* was built in 1507 and demolished between 1592 and 1605. B.M. Apollonj Ghetti, Ferrua, A., Josi, E. and Kirschbaum E., *Esplorazioni sotto la confessione di San Pietro in Vaticano eseguite negli anni 1940–1949*, Citta' del Vaticano: Tipografia Poliglotta Vaticana, 1951.

7   Tiberii Alpharani, *De Basilicae Vaticanae antiquissima et nova structura* [1582], Introduction by Michele Cerrati, Roma: Tipografia Poliglotta Vaticana, 1914, pp. 62, 69, 89, 187.
8   Caradosso's Medal (1505) carries on the verso the inscription "*Templi.Petri.Instavracio. Vaticanus.M[ons]*".
9   Alpharani, *De Basilicae Vaticanae*, op. cit., pp. 3–4.
10 Apollonj Ghetti, op. cit.
11 The author expresses gratitude to Sua Eccellenza Reverendissima Monsignor Vittorio Lanciani, Delegato della Fabbrica di San Pietro, who granted permission to study the 1571 drawing by Alfarano and provided scholarly advice. The author acknowledges Dr Simona Turriziani of the AFSP and Dr Pietro Zander for their precious guidance.
12 Alpharani, *De Basilicae Vaticanae*, op. cit., p. 4.
13 Pierluigi Silvan, "Le Origini della Pianta di Tiberio Alfarano", *Atti della Pontificia Romana Accademia di Archeologia* (Rendiconti Vol. LXII 1989–1990), Tipografia vaticana 1992: 18, pp. 3–25.
14 *Ibid.* Alfarano did not account for the slightly diverging orientations of the old and new basilicas. The west–east axis of the old is rotated two degrees counter-clockwise with respect to the new.
15 *Ibid.* pp. 3–25.
16 Alpharani, *De Basilicae Vaticanae*, op. cit., p. xxvii.
17 Professor Nazareno Gabrielli observed that Alfarano used gold paint (12 June 2009). Information regarding the materiality of the drawing comes from direct observation by the author and from *Notiziario Mensile della Basilica di San Pietro* VI, 1 (Gennaio 1994); Silvan, "Le Origini della Pianta …", *Atti della Pontificia Romana Accademia di Archeologia* 1992: pp. 3–25.
18 Sible de Blaauw, *Cultus et décor, Liturgia e architettura nella Roma tardoantica e medievale: Basilica Salvatoris, Sanctae Mariae, Sancti Petri*, Vatican City: Biblioteca Apostolica Vaticana, 1994.
19 In 1607 demolition of the remaining parts of the old basilica started.
20 Alpharani, *De Basilicae Vaticanae*, op. cit., pp. 25–26.
21 Hans Belting, *Likeness and Presence,* Chicago, IL: University of Chicago Press, 1994.
22 Plato, *Euthyphro. Apology. Crito. Phaedo. Phaedrus*, Harold North Fowler (trans.), Cambridge, MA: Harvard University Press, 1999.
23 *Forma et figura* is translated by Joseph Rykwert as "form" and "appearance". Leon Battista Alberti, *On the Art of Building in Ten Books*, Joseph Rykwert (trans.), Neil Leach, Robert Tavernor, Cambridge, MA: The MIT Press, 1997: 7.
24 By "outline" it is not suggested that the drawing provides merely an indication of the outer edge, rather the proportional elements that constitute an object.
25 These 1590 engravings on copper by Alfarano were commissioned to Natale Bonifacio da Sebenico (1538–1592), and added to the original manuscripts (1582) after 1590.
26 Alpharani, *De Basilicae Vaticanae*, op. cit., pp. 26–27.
27 Barbara Jatta, Director of the Gabinetto delle Stampe of the BAV affirms that the printed garland could be a frame for a coat of arms and that the *bulino* could be the work of the Flemish artist Martin Schoungauer (1448–1491).
28 According to Vitruvius, architecture is composed by "the thing signified" and "that which gives it its significance". Vitruvius, *The Ten Books on Architecture*, New York: Dover Publications, 1960, book I, 1, 3. See Marco Frascari, "Function and Representation in Architecture", in Architecture, 8th Annual Meeting of the Semiotic Society of America; enlarged version in *Design Methods*, vol. 19, n. 1, 1985.
29 Mary Carruthers. *The Craft of Thought. Meditation, Rhetoric, and the Making of Images, 400–1200.* Cambridge: Cambridge University Press, 1998, pp. 102–3.
30 The composite plan was codified by the *quattrocento* architect Francesco di Giorgio Martini (1439–1502), in his *Trattato di Architettura, Ingegneria e Arte Militare*, Milano: Edizione il Polifilo, 1967 [1474–1482].
31 Alpharani, *De Basilicae Vaticanae*, op. cit., p. 26.
32 Apollonj Ghetti, *op. cit.*
33 Alpharani, *De Basilicae Vaticanae*, op. cit., p. xxx.
34 Silvan, "Le Origini della Pianta …", *Atti della Pontificia Romana Accademia di Archeologia*, 1992.

Chapter 14

Neoplatonism at the Accademia di San Luca in Rome

John Hendrix

Introduction

This essay will explore the role that Neoplatonism played in the development of the concept of *disegno* at the Accademia di San Luca in Rome, and its influence on subsequent architecture, for the purpose of illustrating the importance of Renaissance humanism in the culture of baroque Rome. In his treatise, *L'idea de' pittori, scultori ed architetti*, in 1607, Federico Zuccari, the director of the academy, described *disegno interno* as a Platonic idea in the mind of the artist, and a *forma spirituale* which mediates between *archê* and *eidos*, between universal and particular. *Disegno interno* is a *scintilla della divinità*, a spark of the fire of the divine intellect, which is manifest in *nous*, that part of intellect in which divine intelligence participates. Form enters matter through design, and intellect is the instrument which moulds form. For Zuccari, *disegno* is a *segno di dio in noi*, or sign of God in us, as manifest in *intellectus speculativus*, or internal contemplation. Universal design becomes particular design as universals are transformed into particulars, through types, similitudes and metaphors. The spark in *nous* creates dream images, phantasms, and imaginations, along with artistic designs.

Plato, in the *Republic*, distinguished between imitative representation and cognitive process. The artist invests form with concept. Aristotle, in the *Metaphysics*, described the form of the work of art as pre-existing in the mind of the artist. Plotinus, in the *Enneads*, defined form as that which enters matter through intellect. The artist reproduces the principles of nature rather than its appearance. Form or *eidos*, in both art and nature, is produced by the ordering principles of intellect, or *nous*. Marsilio Ficino, in *Sopra lo amore*, or *Commentary on the Symposium of Plato*, written in 1444, distinguished between *bellezza*, sensible beauty, and the *bello*, *disegno interno*, of which *bellezza* is a product. *Res extensa*, external form, is a realization of *res cogitans*, ideas. In Ficino's *Theologia Platonica*, design is seen as a mirror to the intellect, reasoned through "metaphor and similitude". This concept is developed in the transcripts of the Accademia di San Luca in Rome, *Origine et progresso dell' Academia del Disegno*, written in 1604 by Romano Alberti. The distinction between *bello* and *bellezza* is developed by Pietro da Cortona in his treatise *Trattato della pittura e scultura*, written in 1652 with Domenico Ottonelli, using the pseudonyms Britio Prenetteri and Odomenigico Lelonotti.

Pietro da Cortona played an active role in the academy, as did Carlo Maderno, and the influence of the concept of *disegno* can be seen in particular in the architecture of Pietro and Francesco Borromini, in works such as Santi Luca e Martina and San Carlo alle Quattro Fontane, which can be seen as visual and structural representations of the intellect of the artist and the principles of nature. This can be seen in mathematical, geometrical, and metaphorical models of the structure of the cosmos as used by both architects. Precedents for the role of "metaphor and similitude" in architectural design can be seen in works such as Giulio Romano's Palazzo del Tè and Michelangelo's Porta Pia in the sixteenth century.

Zuccari expressed in *L'idea de' pittori, scultori ed architetti*, in the Neoplatonic tradition, that the form of the work of art exists first in the mind of the artist, considering design universally as the fabrication of every intellectual idea.[1] *Disegno interno* is "a concept formed in our mind, that enables us explicitly and

Figure 14.1
Francesco Borromini. San Carlo alle Quattro Fontane

Figure 14.2
Giulio Romano. Palazzo del Tè

Figure 14.3
Michelangelo. Porta Pia

clearly to recognize anything, whatever it may be, and to operate practically in conformance with the thing intended".[2] The idea is a *forma spirituale* which is both "devised and used by the intellect to apprehend all natural things clearly and distinctly",[3] and the expression of such apprehension in artistic materials; here the idea is differentiated from the theological idea as it is applied to artistic production, or artistic representation, which is called *disegno esterno*, or external design. Following Plato, *disegno esterno* is the imitation of an idol, and it is also a metaphor for *disegno interno*, the inner ordering principle and Neoplatonic light of the intellect.[4]

Plato conceived of the idea as being "in the world of shapes and figures something perfect and sublime, to which imagined form those objects not accessible to sensory perception can be related by way of imitation",[5] in the words of Cicero in the *de Orator* (On the Orator). For Plato the Idea, the form of a thing, exists "eternally, being contained in our reason and our intellect: all else is born and dies, remains in a state of flux". As such Plato distinguished between imitative representation,[6] or the ability to render only the sensory appearances of the material world and form which is invested with the concept of cognitive truth in the attainment of a universal and eternal form corresponding to the Idea. Such a distinction would mean a distinction and separation, and also identification, between the aesthetic and the theoretical, between sensory experience and rational cognition.

Form and concept, image and idea should be combined in the work of art. The work of art is defined as being invested with scientific and mathematical principles which it can reveal, as in architecture and music, but can only exist in the realm of images and not ideas, and can thus inhibit as much as aid the knowledge of the Idea – the unrepresentable should not be represented, though form should suggest concept. As the reception of material objects by sensory perception is conceived as a cognitive function, so form should be invested with cognitive process, but can only be done so by that process. Aristotle distinguished between form and matter, appearance and idea as well, but he saw them as interacting, in that form enters into matter, as opposed to being a separate archetype. In Book VII of the *Metaphysics*, "The thing in the sense of form or substance we do not make, but the so-called compound whole we make; and in everything that is made matter is present and one part of it is matter and the other form."[7]

Art, for Aristotle, is a form of the power of thought, and the form of the material product contains the essence, or is the essence, of the form in the idea of the maker. As form and matter interact, so thought and substance interact in form. The form of a work of art exists in the mind of the artist before being translated into matter, and, as for Plato, pure form is indivisible, in contrast to its material existence as part of matter, which is divisible and temporal. In *Metaphysics*, "That which is a union of form and matter can dissolve into material parts; but that which is not bound up with matter, cannot dissolve."[8] Matter is that which springs from the prime unity. It is the single substance to which all bodies can be reduced, the material substrate which is without form.

For Aristotle, the idea is the combination of matter and form. Cognition, as a universal act, an act of the universal intelligence, of which human intelligence is a part, is transformed into the form of the material object or work of art. The work of art is invested with the universal intelligence, unified in the absolute, but dissolved into material parts through individual cognitive acts, then combined in the compound whole. The cognitive acts, the thoughts, are then identified with the material parts. From the *Metaphysics*, "The mind thinks of itself, when it takes on the nature of an object of thought. It becomes an object of thought through its perceiving and thinking, and then thought and object of thought are the same."[9] Matter is integrated with form, and thought is conflated with the object of thought in its actualizing potential as a product of the diversification of the absolute, through the realization of form, as in art. Such a concept would lead to the development of *disegno* and the idea in sixteenth-century Italy, where art is seen as the manifestation of conceptualization itself through the idea, which can be read in seventeenth-century Roman baroque architecture. As Zuccari explains, in *L'idea de' pittori, scultori ed architetti*, "Design is formed form, and imaged in the soul, and in the intellect; therefore our soul and intellect should be as instruments; the senses bring objects, and sensible forms from the intellective part of the soul, and the intellect takes the forms from the forms, that is Design is formed from the same types, and so learns, intends, and knows."[10]

For Aristotle, the interaction between form and matter and between appearance and idea is paralleled by the interaction between the general concept (relating to universal intelligence) and the particular idea (relating to the cognitive act). The appearance in art does not imitate the idea – instead form enters into matter. In that the forms of art can exist in the mind of the artist beforehand, they are different from the forms of nature, but they are like the forms of nature in the interaction of form and matter. In *Metaphysics*, Aristotle illustrates the pre-existence of form in the mind with the conception of the house by the architect and the statue by the sculptor, though the house and statue can only pre-exist as form through the Idea, and not as matter.

The Neoplatonic Idea: Plotinus, Aquinas and Ficino

Plotinus, who taught philosophy in Rome from 245 to 270, separated form from matter as well. He explained, in "On the Intellectual Beauty", the Eighth Tractate of the fifth *Ennead*, "The form is not in the material; it is in the designer before ever it enters the stone; and the artificer holds it not by his equipment of eyes and hands but by his participation in his art" (*Enneads* V.8.1),[11] that is, in the Idea, the conceptualization of the form. As for Plato, form is transferred from the universal intellect into matter, but in a contaminated state, disseminated and no longer ideal. Beauty in art is only a derivation of beauty in the idea or intellect, as a shadow or fragmentation.

The form in the material cannot be compared to the form in the idea. Art, in the conceptualization of the universal intellect through which the form is given, must contain within it an idea of absolute form in the material, an internal idea which must be communicated although its material existence, its external form, falls short.

Plotinus explained in *The Enneads*, that the idea or reason principle must contain a higher degree of beauty and purity than anything produced as art, as an external object. The order of the universe may be revealed through a particular form, as the macrocosm is revealed through the microcosm, through the idea, through conceptualization in form. For Plotinus, this occurs through the intellect or principle which is found in nature. "Artists do not simply reproduce the visible, but they go back to the principles in which nature itself had found its origin" (V.8.1).[12] The generation of the principles of the origin of nature constitute the Idea. The architecture of Borromini, as at San Carlo alle Quattro Fontane, can be seen as a visual and structural representation of the principles of nature as given by Neoplatonic philosophy. The architecture of Leon Battista Alberti, as at Santa Maria Novella, sought as well to reproduce the principles of nature in mathematical and geometric abstractions.

Then the task for Plotinus is to discover what in the principles of nature can be transferred as form in the idea. The matter of nature alone is ugliness without an intellect, an organizing principle, or a superimposed Idea relating the order of nature to the human intellect. Using the metaphor of architecture to apply the principles of nature to art, as constructed from natural material, Plotinus asks, "How can the architect adjust the externally apparent house to the internal *eidos* of the house and insist that it is beautiful? Only for this reason, that the external house, if the stones are imagined away, *is* the internal *eidos*, divided of course with regard to the mass of matter, but indivisible in essence, even though appearing in multiple form" (*Enneads* I.6.3).[13] The work of art is the Idea, the intellection of the internal principles of nature. Structures belonging to appearances are imitations of the Idea, as in architecture, rather than form, as form is divisible in matter but indivisible in the Idea. The Idea, that which is beyond sensory images in the ordering of the universe, is continually present

Figure 14.4
Leon Battista Alberti. Santa Maria Novella

but inaccessible, as it is concealed in the architecture of Borromini, in mathematical and geometrical structures underlying the formal arrangements.

Thomas Aquinas, who lectured in Rome from 1265 to 1267, and began the *Summa Theologica* at Santa Sabina on the Aventine, formulated, in the tradition of Plato, Aristotle and Plotinus, a conception of the Idea, correlating idea with form, beginning with the separation of form and matter, and the idea of form as principle in the universal intellect. As he explained in the *Summa Theologica*, "the Greek word *Idea* is in Latin *Forma*. Hence by ideas are understood the forms of things, existing apart from the things themselves". The form of something, "existing apart from the thing itself, can be for one of two ends; either to be the exemplar of that of which it is called the form, or to be the principle of the knowledge of that thing, according as the forms of knowable things are said to be in him who knows them".[14]

Form is given by the inner principle, the human intellect as a microcosm of the universal intellect. Works of art are differentiated from works of nature in that the form of the man-made object pre-exists in the intellect, though in both nature and the man-made object, the likeness of the form exists in the actuating principle, the creative *nous* or the inner principle, as in the structure of the house, from Plotinus and Aristotle, the pre-existent idea in the *nous* of the architect. In all things not generated by chance, the form must be the end of any generation whatsoever. An agent of generation does not act for the sake of the form, but the likeness of the form is in the agent. In some agents the form of the thing pre-exists according to its natural being. In other agents the form of the thing pre-exists according to intelligible being. Thus the likeness of a house pre-exists in the mind of the architect.

For Marsilio Ficino and the Florentine Platonic Academy, the Idea entails conceptualization in the human intellect, and is no longer a metaphysical substance existing outside the world of sensory appearances or the human intellect. Art is the primary manifestation of the Idea, despite the Platonic restrictions of form as appearance and imitation. Using the example of the architect, Ficino conceived of the process of abstraction in artistic production. He explained in *Sopra lo amore* that "the Architect conceives in his spirit the reason and approximately the idea of the edifice; he then makes the house (according to his ability) in the way in which he has decided in his mind". He then asks, "Who will deny that the house is a body, and that it is very similar to the incorporeal Idea of the artisan, in whose image it has been made? It must certainly be judged for a certain incorporeal order rather than for its matter."[15]

The reconstruction of the image of the house or body in the imagination is given by the inner ordering principle of the sensory images. The order or form is incorporeal, in abstracted matter, as opposed to the matter. The order, emanating from the universal intellect, is preserved between the corporeal and incorporeal, and the order or form can exist without body or matter, as it is prior to it. Along with the abstraction of form and inner sensory images comes the Neoplatonic emphasis on interior illumination in the processes of imitation, as in the ordering of the rational principles of nature in Alberti's *concinnitas* and the establishment of canons of beauty or *bellezza*. Unlike Plotinus, form is a product of the human intellect while matter is a product of the exterior world, and *res cogitans*, or things thought, are separate in their generation from *res extensa,* or things external, though connected by the inner ordering principle.

Bellezza, or sensible beauty, and the realization of the *bello*, the idea of the beautiful, in artistic representation, are manifestations of the metaphysical existence of form as well in Neoplatonic thought, given a priori in the ordering principle of the intellectual vision of form, as opposed to matter, and as opposed to generation and synthesis in the phenomenological congregation of matter. In the metaphysic of the *bello*, aesthetic value, representative of moral value, is transferred in the ordering principle, creating interior illumination and the capacity to transfer the *bello* to works of art. *Bello* is extracted from *convinto*, or conviction, but *convinto* is not associated with the generation of forms by Ficino, although it is by Gian Paolo Lomazzo in the *Trattato dell'arte della pittura*, published in 1584.[16] Lomazzo combined the metaphysical idea of the *bello* with the process of the artistic conception in Aristotelian terms; the metaphysic of interior illumination is combined with the interaction of matter and form in universal cognition and the making of things.

Disegno interno at the Accademia di San Luca

Disegno interno exists a priori of artistic execution and is combined with Christian theology by Zuccari in being the manifestation of the universal intellect, or a *scintilla della divinità*. The Idea is implanted by God in angels without sensory perception, and then it is transferred to human intelligence, as the aspect of universal intelligence in human intelligence. Works of art are meant to be thought as works of nature in resemblance and correspondence, in embodying the same internal principles, as in Leon Battista Alberti's *concinnitas*. Following Plotinus, the forms produced or replicated from an incorruptible substance are varied, infinite, fragmented, corrupted and disseminated, as in Borromini's architecture, in comparison to the pure and unified forms of the absolute, as in the lantern of San Carlo, in the pure light from the oculus and the resolution of architectural forms. Baroque forms contrast the intertwined and variegated corporeal forms of the terrestrial realm in Neoplatonic hierarchies to the pure forms of the absolute, representing a passage through the tortured labyrinth of reality. For Zuccari, forms of design can only be the result of accident, as the artist "forms within himself various designs corresponding to the different things he conceives. Therefore his design is an accident, and moreover it has a lower origin" than the design of God, "namely in the senses".[17]

Human designs are formed by a process which Zuccari compares to a spark, as in the *scintilla della divinità*, the spark being the first concept created from the intellective virtue, or Idea, the first manifestation, in Aristotelian terms, of the Idea in the sensory imagination, being, as in an accident, indeterminate and confused. The spark is the transformation of the Idea, as matter and form, into cognition, retaining both matter and form, following Aristotle. The light from the spark is interior illumination, which lights the senses and the intellect, and is disseminated in variegated material, as in San Carlo and sculptural ensembles such as Bernini's *Ecstasy of St. Theresa*. Order is manifest in material form, being transferred through the ordering principle of the Idea. In *L'idea de' pittori, scultori ed architetti*: "[T]he intellective virtue strikes the stone of concepts in the human mind, and the first concept, a spark ascending the tinder from the imagination, and moves phantasms, and ideal imaginations", while "this first concept is indeterminate and confused, born in the faculty of the soul, or the agent of the intellect possible and understood."[18]

Disegno interno is a sign of the possibility but imperfection of human knowledge. Nature achieves order through an intellective principle, formulated through the projection of human logic and reason onto the mute forms of nature. Art imitates nature in so far as it enacts human reason, of which the forms of nature are representative: "Nature is guided toward its own goal and toward its own procedures by an intellective principle." Since "art, chiefly with the aid of the above-mentioned design, observes precisely the same in its procedure, therefore Nature can be imitated by art, and art is able to imitate Nature."[19]

Because the forms of nature are combinations of form and matter, art is capable of forming itself from the Idea or intellect. Through the agent of the mind of the artist, artistic forms arise as natural forms, forms being transformed into matter, accidental and discernible, revealing in the indeterminateness and confusion of sensory experience, the clear order of the intellective principle. The multiple and variable forms of the material world emanate from the unity of the absolute as the multiple forms of baroque architecture in the worship space of the church emanate from the unified and resolved source of light in the lantern. Multiplicity and variability are the result of the limitations of the human intellect. Following Aristotle, *disegno interno* is a process of actualization, and intellect is an operating instrument. *Disegno* involves the phantasm, formed concept and object of sensation, a mental representation of a real object, a product of the imagination, a manifestation and reflection of form according to the intellective principle. "Design is form, concept, idea, and light of the intellect, and in its potential of intention and understanding, its first intellections ... are not without phantasms; phantasms, say the philosophers, are sensible things, ideally formed concepts."[20] *Disegno* also entails dreams, which serve to form, in the internal sense of the imagination, prototypes for representations of form in images, through cognition and judgment. "Imagination is formed by common sense in cognition, judgment, and the division of types, composed together as we experiment in dreams, having seen [objects of perception] represented in dreams."[21]

For Zuccari, "[W]henever our intellect forms within itself some Universal Design, at the same time the two internal senses [cognition and imagination] form together their own particular design."[22] The *disegno esterno* of sensory experience is the particular correlation to the universal *disegno interno* of abstraction and intellection in the Idea, the particularity in matter as defined temporally of universal form. The formation of particulars in *disegno esterno*, and the identification of physical types, as in dreams, is a scientific process: "Our intellect has in this way begun to intend something, and from that [process] science is acquired, and after being placed in it some type is acquired, and science itself, also forms in itself various representative designs of this particular first known through the senses."[23] Zuccari explains that the intellect first knows the nature of material things given by the internal principle, then it knows the types in which the nature is reflected, then it knows the singular forms caused by the types. The forms are known by the two internal concepts, sensed and unsensed, particular and universal, formed from imagination and cognition,[24] and become a particular design representing the singulars of the particular nature.

Design is the means by which the Idea in the intellect becomes the external form of the artistic production. Design is "both intellective and practical, corresponding to two intellects in us, the speculative one for universal purposes, and the practical one for the purposes of our operations",[25] as described in the transcripts of Romano Alberti. Both types of design are necessary, "as one is objective and has known limits in practical and particular things and the other represents things which are universally understood in the intellect. Thus there is an intellective design and a practical design."[26] Design as a whole is the "expression and declaration of a concept, which is first in the mind".[27] *Disegno interno* fabricates the Idea, while *disegno esterno* fabricates the exterior visual experience. *Disegno esterno* is "defined in visual form. It is the form of all exemplary forms of all things which we can imagine and form".[28]

External design "is a form of science, a faculty for determining the proportions of quantity in visible things",[29] a faculty assigned by Marsilio Ficino to the mechanism of vision in the *Theologia Platonica*. But external design is also "a definition of our intelligence, reflecting the intellect in a clear mirror, making visible the things represented in the intellect, through intelligible forms, reasoned in metaphor and similitude".[30] Artistic forms come about through the process of design in the use of literary tropes from Aristotle's *Rhetoric*, including metaphor and similitude. Meaning in the visual arts is given by language, as visual forms are the manifestation of ideas.

Notes

1. Federico Zuccari, *L'idea de' pittori, scultori e architetti*, Torino, 1607, in *Scritti d' arte di Federico Zuccaro*, Detlef Heikamp (ed.), Firenze: Leo S. Olschki Editore, 1941, p. 95.
2. Translated in Erwin Panofsky, *Idea: A Concept in Art Theory*, Columbia, SC: University of South Carolina Press, 1968, p. 85.
3. Panofsky, *Idea: A Concept in Art Theory*, p. 86.
4. Zuccari, *L'idea de' pittori, scultori e architetti*, p. 95: "Il dissegno considerato particolarmente, in quanto che è parte, è fondamento della Pittura, Scultura e Architettura e imitazione d'un idolo che consiste nelle proportioni di quantita continua, perfettamente disposte, e determinate, formato dall' istess' Artefice. Questo Dissegno nell'istesso modo considerato, si puo deserivere per metafora, che sia luce dell'intelletto, e vita dell'operationi, o vero realmente che sia faculta di

determinare perfettamente le proportioni di quantita, nelle cose visibili. Considerando il Dissegno universalmente, in quanto che alla fabricatione d'ogni Idea intellettuale, si può quel nome applicare … cioè che sia lume generali dell'intelletto; Primo motore dell'intelligentie; forma de tutte le forme; Alimento delle pratiche; un'altro Nume, o Natura generante, che aviva, opera, e da spirito a tutte l'intelligentie humane".

5   Translated in Panofsky, *Idea: A Concept in Art Theory*, p. 11.
6   Panofsky, *Idea: A Concept in Art Theory*, p. 3.
7   Translated in Aristotle, *Metaphysics*, Louise Loomis (ed.), New York: Walter Black, 1953, p. 29.
8   Translated in Panofsky, *Idea: A Concept in Art Theory*, p. 27.
9   Translated in Aristotle, *Metaphysics*, p. 35.
10 Zuccari, *L'idea de' pittori, scultori e architetti*, p. 37: "Il Disegno è forma formata, e imaginata nell'anima, e nell'intelletto, e pero vuole che l'anima nostra, e l'intelletto sia nel sudetto modo quasi un'istessa cosa e come mano che piglia la forma delle forme, cioè il Disegno formato dall'istessa specie, e cosi apprende, intende, e conosce."
11 Translated in Plotinus, *The Six Enneads*, Stephen MacKenna and B.S. Page (trans.), Chicago, IL: Encyclopedia Brittanica, Inc., 1952, p. 239.
12 Translated in Panofsky, *Idea: A Concept in Art Theory*, p. 26.
13 Panofsky, *Idea: A Concept in Art Theory*, p. 28.
14 Translated in *The Basic Writings of Saint Thomas Aquinas*, Anton Pegis (ed.), New York: Random House, 1945, p. 162.
15 Marsilio Ficino, *Sopra lo amore o ver convito di Platone*, Florence, 1544, Or. V, Ch. 3–6, pp. 94ff, in *Opera latina*, II, 1336ff (translated in Panofsky, *Idea: A Concept in Art Theory*, p. 137).
16 See Edmondo Cione, "Presentazione del Traduttore", in Panofsky, *Idea: A Concept in Art Theory*, p. xv.
17 Zuccari, *L'idea de' pittori, scultori e architetti*, pp. I, 7, 50, translated in Panofsky, *Idea: A Concept in Art Theory*, p. 88.
18 Zuccari, *L'idea de' pittori, scultori e architetti*, p. 25: "Pero dico, che si come per formare il fuoco il focile batte la pietra, dalle Pietra n'escon faville, le faville accendon l'esca; poi appressandosi all'esca i sosarelli s'accende la lucerna; Cosi la virtu intelletiva batte la pietra de i concetti nella mente humana; e il primo concetto, che svavilla accende l'esca dell'imaginatione, e move i fantasmi, e imaginationi ideali, il qual primo concetto è indeterminato, e confuso, ne dalla facolta dell'anima, o intelletto agente possibile e inteso".
19 *Ibid.*, p. 21: "La ragione poi, perche l'arte imiti la Natura è; perche il Disegno interno artificiale, e l'arte istessa si muovono ad operare nella speditione delle cose artificiali al modo, che opera a Natura istessa. E se vogliamo anco sapere perche la Natura sia imitabile, e perche la Natura è ordinata da un principio intelletivo al suo proprio fine, e alle sue operationi; onde l'opera sua è opera dell'intelligenza non errante, come dicono i Filosofi".
20 *Ibid.*, p. 37: "Dunque il Disegno è forma, è concetto, è Idea, è luce all'intelletto, e alle potenze sue all'intendere, e al capire, e le prime intellettioni non sono fantasmati, ma non sono senza fantasmati; perioche i fantasmati, come il Filosofo vuole, sono cose sensibil, concetti formati Ideali".
21 *Ibid.*, p. 26: "Il secondo senso interno si chiama fantasia, è formate dal senso commune nella cognitione, giuditio, e coparatione di quelle specie, riceve in se stesso, e in oltre queste insieme compone, come noi esperimentiamo ne i sogni, ch' havendo veduto".
22 *Ibid.*, p. 31: "che mai l'intelletto nostro forma entro di se qualche Disegno universale, se anco prima, o insieme questi due sensi interni non formano i proprii Disegni particolari".
23 *Ibid.*, p. 32: "L'intelletto nostro ha in questo modo cominciato ad intendere qualche cosa, e di lei acquistata la scienza, o pure doppo posto in esso qualche specie acquistare, e la scienza istessa, forma anco in se stesso varii Disegni rappresentanti quel particolare conosciuto per il senso".
24 *Ibid.*: "In prima egli dirittamente conosce solo la natura commune delle cose, e poi indirrittamente, e quasi reflettendosi sopra di se medesimo, considerando, e la specie, con la quale primo conobbe, e l'operatione con la quale conobbe, conosce anco il singolare dal quale detta specie in qualche modo fu cagionata; Cosi non potendo lui in alcun modo, o compitamente, o incopitamente intendere senza Disegno, forma anche di quel singolare un Disegno interno in se stesso. Si che intendono noi qual si voglia cosa; prima la fantasia, e la cognitione dei singolari di quella formano due concetti, uno sensato; e l'altro insensato, poi l'intelletto dirittamente intendono,

forma il Disegno interno commune rappresentante quella cosa in generale, quanto alla natura comune, in ultimo indirittamente discorrendo, ne forma un'altro Disegno particolare rappresentando i singolari di quella natura".

25 Romano Alberti, *Origine e progresso dell'Academia del Disegno*, Pavia: Pietro Bartoli, 1604, p. 19: "Si deve pero sapere che non d'una, ma di due forti, è il disegno, cioè intellettivo, e prattico, poiche si come sono duoi intelletti in noi uno chiamato speculativo, il cui fine proprio è l'intendere solamente in universale, e l'altro adimandato intelletto pratico, il cui termine proprio, e ultimo è l'operare, o per dire meglio esser principio dell'operationi nostre".

26 *Ibid.*: "è necessario che anco siano due i disegni alluminanti gl'intelletti nostri, cioè uno che è oggetto, e termino dell'intelletto cognoscitivo, e questo rappresenta all'intelletto le cose universalmente intese, e l'altro che è oggetto e termino dell'intelletto prattico, e questo rappresenta all'intelletto le cose in particolare, e in singolare".

27 *Ibid.*, p. 18: "Il disegno è una apparente espressione e dichiaratione del cocetto, che era prima nell'animo".

28 *Ibid.*, p. 19: "che cosa sia il disegno esterno, in forma sua visiva … hora il pratico dico essere forma di tutte le forme essemplare di tutte le cose, ch'immaginare, e formare si possono".

29 *Ibid.*, p. 18: "la forma della scienza … che dice il disegno essere una facoltà di terminare perfettamente le proportioni di quantità nelle cose visibile".

30 *Ibid.*: "è insieme un termine della nostra intelligenza, in cui come in lucidissimo specchio l'intelletto chiaramente, e espressamente vede le cose rappresentate in lui, per le forme intelligibili ornanti l'istesso intelletto, o ragionando in metafora, e similitudine lo difiniremo".

Chapter 15

Who's on first?

Donald Kunze

The defense of the humanities

Costello: Well then who's on first?
Abbott: Yes.
Costello: I mean the fellow's name.
Abbott: Who.
Costello: The guy on first.
Abbott: Who.[1]

From Percy Bysshe Shelley's "Defense of Poetry" (1821) through C.P. Snow's "Two Cultures" (1959), the humanities' classic rhetorical stance has been one of defense in the face of science. This defensiveness has marginalized the arts, but even more damaging is the assumption that natural sciences are precise while the human sciences are generalistic. But, in contrast to the objects of nature – by definition alien from the theories that attempt to explain them – human objects and actions are theoretically open to full and complete understanding, even if this understanding must adopt specific strategies and forms in the face of the paradoxes of self-reference ("we are what we study").

Giambattista Vico (1668–1744) was the first thinker to realize the implications of this through his motto of *verum ipsum factum* ("humans may in theory know what they have made"). Precision of human sciences is a function of the "precision" of subjective structure, whose invariance affords cultures' and history's infinite diversification. Later, Jacques Lacan (1901–1981) qualified this precision as based in language, but in a language that confronted the human subject with an identity that did not exactly fit. For both Vico and Lacan, a metonymy of inside for outside puts the human mind *outside of itself*. As Agent Mulder said in *The X-Files*, "The truth is out there!" The truth *and its dislocation* were the basis for both Vico's "optical" construction of subjective universality and Lacan's idea of the "extimate" (*extimité*) in the famous "mirror stage." Because dislocation is an *architecture* in the "optical identities" established by Vico and Lacan, architecture occupies a privileged position within the humanistic theory of both thinkers and, indeed, anyone who attempts a theory of subjectivity.

The mirror stage's place-as-temporality

In her lucid and witty review of Lacan's mirror stage, the scholar Jane Gallop couples the bibliographical problem of determining which of Lacan's articles is historically the "first" written account of his oral and unrecorded presentation in 1937 with the issue of what comes first in the mirror stage itself.[2] The scholar's predicament of finding a "primary text" to serve as the source for the idea as it first emerged in print is, in an uncanny sense, a mirror of the very subject the scholar has isolated for study. Sandwiched in between is a greater, more breathtaking parallel, the realization that just as the mirror stage is the beginning of the child's anticipation of control based on an image of unity, Lacan's own beginning as a thinker begins with this idea, which is in many ways structured exactly like the mirror stage.

Where have we encountered this kind of multiple mirroring before? Thanks to the comparative obscurity of this eighteenth-century Neapolitan philosopher of culture, only a few readers might supply the answer that I believe fits best: Giambattista Vico's *The New Science*, where the author places, as the origin of not only his object of study but his awareness of the "science" of that object, the "imaginative universal" (*universale fantastico*). Vico's own insights begin with the historically remote moment that initiated human thought proper: a moment when humans attribute their own qualities to the sky. Thunder occasioned this transfer though a metonymic sequence of sounds that the first humans took to be the voice of the sky.[3] Soon, all of nature was animated by human-like character and intentionality. If this idea is the generative origin of the entire *New Science*, the question is, how did Vico think of it?

Vico gives an account of his discovery principle in terms of a special universal invented at the opposite end of history from the myth it is empowered to discover: *recollective fantasia*, a combination of practices he elaborates throughout his earlier works and *Autobiography*. This form of *fantasia* is for the modern scholar, the modern reader; anyone who would aspire to understand the remote, strange thought of the first humans. At the end of history, this combination of memory and imagination is a form of and counterpart to the very thing it attempts to discover: a key that both mirrors and deciphers the imaginative universal. But, this form of mastery, like Lacan's mirror stage, promises future success that retroactively fragments the scholar's "past." The imaginative universal and the scholarly universal required to unlock it are antipodes of a globe that requires clever navigation, a trip that cannot penetrate directly through but must sail around.

If we take Vico's late work on the scholarly universal in combination with his personal account of intellectual development, we have an interesting story.

Figure 15.1
Otto Vænius (1556–1629). "Cebes' Table." Source: *Theatro Moral de la Vida Humana*, 1672. By comparing the table to a *mons delectus*, or "mountain of choice," Vænius is able to deploy Vico's ideas of the *cœlum*, heaven as mind (*animus*), able to penetrate mute matter (the clouds). At the same time, Vænius demonstrates the primary architectural function of the table: a distinction between the otherwise "anamorphic" building types, the temple and the labyrinth.

Working in his father's bookstore, the young Vico fell off a ladder and sustained a concussion. The attending physician predicted that the boy would either die or grow up to be an idiot. Vico leaves it for the reader to disregard the physician's prediction or take it up in a new light – that Vico's highly original and widely misunderstood thought did in fact amount to a death narrative.[4] Vico follows up this joke with a series of images of dire alternatives. At one point, he reports that he did not know whether he, in his search for a totalizing theory, was a "god or devil." He employs a mystical-esoteric image as a frontispiece to the last edition of *The New Science*, based on "Cebes' Table," a text that was well known in Vico's day. The story accompanying the table is interesting. Pilgrims visiting the Temple of Saturn notice an image in the shadows at the rear of the temple and ask the attending priest about it. It is an image of wisdom, he replies, but be careful if you wish to look at it. If you understand it, you will be transformed by its wisdom; if you fail, you will go mad. Some of the many versions of this text propose an image that approximates the painting in the back of the temple: a labyrinth surmounted by or surrounding a

Figure 15.2
Domenico Antonio Vaccaro (1678-1745). "La Dipintura Allegorica."
Source: Giambattista Vico, *The New Science* (Venice: Felice Mosca, 1730). Vico's "memory place" and mythic clearing in the Nemean wood, first printed as the frontispiece of the second edition of *The New Science*.

temple; or, as in some versions, a *mons delectus*, a steep pathway leading to a pinnacle occupied by a temple above the clouds. Again, the high-stakes allure of discovering the "first human idea" pervades not just Vico's advice to the reader of *The New Science*, but also the idea that initiates that work itself: the thunder that shocked the first humans into thinking like humans, and the idea that, once discovered as a principle, grounded the science of that moment and its aftermath.[5]

It is curious that both Lacan's and Vico's generative insights involve mirrors in literal as well as figurative ways. In Lacan, it is after all the mirror that initiates the child's (mistaken) anticipation of unification based on the specular image. Vico makes something of the same comparison. In the image that serves as the frontispiece of *The New Science*, known as the *dipintura*, a divine eye enclosed by a triangle casts its gaze onto a reflecting jewel on the breast of Metafisica, who surmounts a globe representing the limits of the perceptible world (Figure 15.2). The visual ray reflects onto a statue of Homer, as frozen as the victims of Medusa and himself famously blind. "Blind to what?" would be Lacan's question, and one might say "blind – and indifferent – to the objects that normally impede human perceptions and actions, the Lacanian 'part objects' that mark the limit of human attempts at mastery, such as the gaze and the voice."[6]

Objects lying at Homer's statue's feet metaphorically defend the imaginary and symbolic realm: scales representing justice, a purse for commerce, fasces for centralized government. These represent the accretions of the collective cultural ego: the *certum* or concretizations produced to protect from imagined–real lacks/threats, which, like the plow and tiller further into the image, develop humans' heroic mastery of nature. The limits of perception that compare to Lacan's part objects are on the altar and the globe, precariously placed at the corner of the altar: the rituals of fire and water, used to sanction the boundary crossed at both marriage and burial, which, Vico notes, were originally the same ceremony.

Standing before the mirror of Metafisica (nature made into a mirror), the first humans "misrecognize" themselves by seeing the bodies and actions of gods. This is not the same as the human infant seeing a mirrored self-image, but the parallels are nonetheless instructive. The (at first unrecognized) act of mirroring is what Vico credits as the essence of his insight. It is the metaphorical capacity of the mind to conceal itself from itself, and this capacity is a quick succession of the very powers of metaphor that Lacan himself cites. The thunder is essentially metonymic. It does not say anything that the first humans can understand. Rather, it creates a "meaning effect." The author and Vichian enthusiast James Joyce speculated that it contained the necessary phonemic components of all languages, just as linguists note that the infant, in babbling, will run through all vocal possibilities and gradually discard the ones not used as he or she learns a native tongue. Metonymy, Lacan devises, is a means of signifying through absence.

$F(S...S') \cong S(-) s$

Ed Pluth provides an expert translation:

> The Ss stand for signifiers, and the *s* for a signified effect. This formula expresses much that we already know about metonymy: the movement from one signifier to another in the signifying chain (S…S') is congruent to or tantamount to (S) one signifier giving the effect of there being a signified somewhere, an effect that is not placed in the signifying chain but that "resonates" beyond the signifying chain, indeed, beyond the signifier itself (S–*s*). The bar between S and *s* can then be taken to represent a gap between signifiers and the signified effect but also as a minus sign, such that metonymy gives us signifiers with an absent signified effect. "Resonance" is perhaps the ideal term for expressing what it is that metonymy achieves.[7]

Through this formula, we can see the importance of metonymy to the mirror stage. Gallop notes:

> The mirror stage is a decisive moment. Not only does the self issue from the mirror stage, but so does "the body in bits and pieces." This moment is the source *not only for what follows but also for what precedes*. It produces the future through anticipation and the past through retroaction. And yet it is itself a moment of self-delusion, of captivation by an illusory image. Both future and past are thus rooted in an illusion.[8] [emphasis mine]

Metonymy's "meaning effect" is created through an absence, a *defect* in the succession of signifiers that, intending to signify, provide only a partial supply. Meaning is created by resonance, and resonance is made possible by the distance constructed between the "partialized" subject-in-bits-and-pieces (the subject that realizes itself as "formerly fragmentary," or *morcelé*). This is a subject made uncanny by a marginalization that invites it into a space but does not allow it properly to belong in that space, to be at home.

Vico's mirror stage

Vico's subject fits perfectly into the Lacanian schema – and in two ways that compound the mystery of how Lacan's mirror stage is the origin of the phenomenal subject *and* a science of the phenomenal subject. Vico's visual evidence is compelling. We have in fact the *subjet morcelé* before our eyes: the pieces of human civilization *as* pieces, in the style of the emblem books of the eighteenth century, cast about the ground of the "oculus" or clearing in the forest, an opening created by the first humans to gain a better view of the provident signs of the sky; past and future, divided by a mirror.[9] Moreover, Metafisica stands on a globe that is in fact an inverted topological model of the heavens, an outside made into an inside, the "beyond" of Elysium contained within the mysterious sphere: a case of Lacan's idea of "the extimate" (intimate exteriority; exterior intimacy), if there ever was one! We see Homer as a statue, not a human figure, as if to emphasize the "permanent erection and stasis" of the materiality of poetry: the dream of unity as a projection in a projected space, appropriately flawed – the base is cracked. All objects in this image are described meticulously in Vico's preface, all but one, that is: the helmet of Hermes. Homer gestures directly at it but Vico doesn't explain its presence. What is the meaning of Vico's silence on this image of invisibility and secret wisdom? When Vico says, just before he sets out his claim to have discovered the key to human construction, that "we must reckon as if there were no books in the world," is he

15.3
Lacan's idea of metonymical meaning follows the structural model of the "anacoluthon," the figure of rhetoric where a terminal element retroactively revises the meaning of the succession of signifiers by altering, through a process of resonance (cf. anamorphosis, parallax, *moiré*) the idea of the initial element, "who's on first."

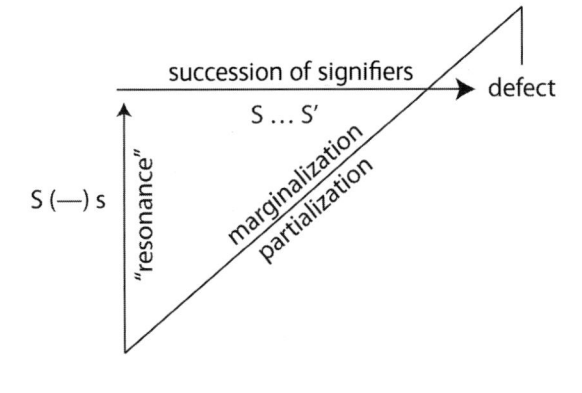

returning to Plato's principle of never writing down any complete thesis but, rather, using language to create a fragment of what the reader–listener must complete silently, internally? Vico's own research was the opposite. According to Verene, he consulted every available text, was an avid reader, a bibliophile.[10] Yet, Vico was too much a Platonist to regard the written word as complete. Even if he had not been familiar with Plato's "Seventh Letter," he would have learned the lesson of *The Dialogues*, that truth cannot be literally stated, that it does not yield to the demands of the intellect to display itself; that it can only be "encountered" through error, dreamlike reflections, and myth. The *dipintura*, and no less the text of *The New Science* itself, should be regarded as *tesseræ*, fragments of a whole that must be "completed by the audience."

To condense what should properly be a lengthy exposition on this subject, consider this shortcut. The literary device of the "unreliable" or "defective narrator" has long served fiction as a means of creating a doubled point of view. With the realization that the narrator is possibly a fictional creation of a "real" author, who intends the reader to deploy an ironic rather than naive understanding of the literal account, a space much like the space of the mirror stage is opened up. The point of view is misrecognized, taken up by a false voice. A "future space" is opened up, where the original single vanishing point of intentional meaning is doubled and competitive. Like the dramatic device of twins who mischievously change places to fool the naive lover, the object of knowledge or resolution of the plot blurs, vibrates, shifts. Once the geodesic point against which all other lines were held to account, the vanishing point is doubled by the unreliable/defective narrator, the voice that undermines order from the inside out.[11]

The space of the defective narrator creates doubles, just as Lacan's mirror creates the child's composite perception of a reflected image, which puts the child into the Lacanian order of the imaginary, and the gaze of the other, which puts the child into the Lacanian order of the symbolic. These two images, left and right versions of the imaginary, constitute a "stereo-gnosis," literally a knowledge of the world through the touch, in this case, of tangent versions of the subject. Like twins, *Doppelgängers*, and rivals, the "minimal difference" between the two nearly identical versions creates a space that can't be crossed, a bomb that can't be disarmed. The stereognostic of the mirror image and the gaze, the image that only the subject can see directly and the image that others "mistake" for the subject, is the necessary and sufficient condition for an inversion that frames space from the inside out. The new space "partializes" the subject in retrospect, makes the narrator into a defective narrator, keeps the point of view "infantilized," fragmented, *morcelé*. The defective narrator is like Humpty Dumpty. He is reduced to a broken child, just as Vico's *Autobiography* tells of his childhood fall from the ladder in his father's bookstore.

Critics of Vico return perpetually to the theme of lack: Vico is difficult to read; he "cannot possibly mean" what he says; he repeats himself. He originates ideas that others develop but seems to have "arrived too early." *The New Science* text is a hodgepodge, a hopscotch: themes and ideas are repeated laboriously.

Only a few critics have suggested that Vico's shortcomings as a writer might actually be intentional. Margherita Frankel argues that *The New Science* was designed as a spiral, where the narrative line repeatedly encountered the same radial theme lines.[12] Vico's *Autobiography* gives away the secret that Vico staged his narrative personality to match the profile of the melancholy genius – between Saturnine paranoia and manic enthusiasm.[13] "Vico did not know whether he was a god or demon," he wrote, mirroring the challenge given in the story about Cebes' Table.[14] That is, although Vico identified his discovery with the image of mastery lying in the future of the mirror's virtual image, he placed his mental turmoil and writing on the side of the fragmented subject strewn in front of the glass. Vico even attached a psychoanalytic explanation to account for his defective narration. Using the classical system of humors, he diagnosed himself as both melancholic and choleric (the eighteenth-century version of manic depression). Rather than crediting Vico as simply a born-too-early clinical psychologist, however, it is important to note that the system of humors connected its psychological applications with history, geography, cosmography, physics, theology, and poetics. The choleric/melancholic link was a code for the heroic poet (or poetic hero), one whose primary field of action was based in event (cf. Lacan's "act") but whose awareness was aligned with the underworld (cf. Lacan's "fantasy").[15] The opposition of choler and melancholy duplicated the mirror image's structured opposition of mastery and failure. Vico's deployment of the phrase *aut Deus aut Demon* ("god or demon") repeated this logic, since the hero, at the "top" of the cycle of humors in terms of power and wit, was also historically credited with being melancholic. Melancholy was not just irony and depression; it was the literal power to investigate Hades and return, the theme of *katabasis*. What was for myth the story of the hero's ability not just to visit the underworld but to resurrect the dead, was in scholarly terms the power of the voice of the underworld as such: the prophetic sibylline insights that can be expressed only through riddles and silences.

Mi-dire (saying half)

The melancholy hero that Vico devised as his literary persona embodied the metonymical subject. The *subjet morcelé* in front of the mirror is, after all, a back projection: a subject cast into the past of a constructed history, Vico's autobiographical self. Just as the metonymical thunder, a series of syllables without meaning, only a "meaning effect," jolted the first humans into humanity proper, motivating them to cut clearings into the forest. The clearing's boundary is what Pluth called "the gap between signifiers and the signified effect but also as a minus sign, such that metonymy gives us signifiers with an absent signified effect." This absent signifying effect was at first acousmatic, as the "voice" embodied in the thunder.[16] The signifiers, as in Vico's *dipintura*, lie strewn about the opening, just as did the pieces of sacrificed animals in early divination rituals, the enactments of the minus sign. The *subjet morcelé* is both the sacrificial victim and the artificially constructed "defective" narrator, a fictional victim or *fictim*, who gives only half the story.

Vico's theatrical organization of *The New Science* makes an interesting prolegomena to Lacan's mirror-stage works. Even the (apocryphal?) story about how the *dipintura* got inserted into the text pages of *The New Science* "at the last minute" recreates the controversy about which of Lacan's texts comes first.[17] Is Lacan's warning about finding later views in antecedent texts more than "just an error of anachronism"? Is it an *apotrope*, a warning to those without eyes to see or ears to hear to turn back immediately – a warning we have already encountered in the case of Cebes' Table? Beyond the *apotrope* is, after all, the domain of the prophetically blind, who are, in the mirror psychology of transitivity, also invisible.[18]

The answer to this question has a central significance for the role of the humanities in architecture and other spatial practices that engage ideas about the space of the subject. In the first place, the central feature of subjectivity becomes the boundary that, as mirror or screen, separates space into two parts with radically different fundamental natures. The subject is, in an equally fundamental sense, the agent who must traverse these two spaces and the act devised to sneak across the border. The subject's crossing, and the intransitivity of the boundary crossed, subsequently color all of the subject's moments – actual and virtual – including the extension of the subject's mastery through visual perception. What is the role of the mirror stage? Is the developmental "fact" of the mirror stage so axiomatic that it was discovered and developed independently some 200 years before Lacan's formulation of it? Do Vico's and Lacan's "versions" constitute antipodal variations on the mirror-stage theme, a kind of "before-Freud" and "after-Freud" dating system that chronologically frames the same kind of "anachronisms" that Lacan set up as apotropaic warnings? Most interestingly, is humanistic theory itself defined and developed by its own subjectification, its construction of its own mirror stage, its own system of apotropaic warnings? Are theory's twin aspirations, for congruence on one hand and disorderly openness on the other, sustained by the mirror stage's forwards and backwards projection towards unity prefaced by fragmentation?

Here, it seems that Lacan has deployed Vico's device of the helmet of Hermes. This is the only object shown in the *dipintura* that is not explained in the text. Hermes is the god not just of commerce but of wisdom, trickery, seduction, and theft.[19] He conducts the souls of the dead to Hades. His power of invisibility is linked to movement and secrecy. In a famous image, Hermes is shown holding a candelabra representing the seven planets and holding a finger to his lips to indicate the secrecy and silence that protect the motion of the soul across this space in the twinned processes of birth and death. Secret motion? One thinks of the labyrinth as puzzled motion, the dances once performed by the Greek chorus, the imposed silence and invisibility of Eurydice as she is led from Hades by Orpheus. Lacan's case is not so esoteric, but his interest in antiquity and use of puzzles, particularly in his reference to Borromeo knots, Klein bottles, and Möbius bands finds a modern scientific and clinical context for restating these matters in their full complexity.

Catherine Clément returns repeatedly to the theme of Lacan's way of speaking. Early in his career, he was fascinated by the speech of the Pepin sisters, two servant women whose brutal murder of their employers in 1949 created a national scandal. The power of the fragmented speech of these women to evoke puzzled but prophetic meanings led Lacan to regard metonymy's ability to generate meaning through absence as primary. Metaphor was, in contrast to metonymy, dependent on the idea of a whole. Metaphor could create meaning by organizing signifiers that had no prior relations. It was a "master signifier," a signifier able to organize other signifiers. Metonymy in contrast depended on the boundary that, in the mirror stage, created meaning effects through remote control and inverted uses. Fragmentation both undermined and reinforced wholeness. It came before and after. It certified its own opposition and destruction. It *was* a boundary, pure and simple, and this boundary was both a dividing line and a minus sign.

The key was Lacan's discovery that his science must "suffer" the same metonymic-paranoiac symptoms as its primary objects of study. Humanistic knowledge of the subject must find a way to collapse the distance and distinction between itself and its materials without losing its ability to make sense. The topological interpretation of this problem was clear. Science had to create an "internal dimensionality," a space and time without recourse to an eternal template of verification. It had to work, in short, with resonance rather than correspondence. The speech of the Pepin sisters and other famous paranoiacs was called *mi-dire*, "half speech." This linked it to the more historical, and less hysterical, forms of prophetic signification: the speech of the oracle, the murmurings of the tribal shaman, the fragmented *tessera* whose reunion constituted a proof of authenticity. These associations reinforced Lacan's already obtuse style of writing and speaking. "I always tell the truth, but I cannot tell all the truth," he was fond of saying.

Conclusion

By inventing an "optical identity," Vico and Lacan show how metonymy accommodates theory's twin obligations, to coherence on one hand and contingency on the other. Itself sustained by the mirror stage's forwards and backwards projection towards unity "prefaced" by fragmentation, theory as *mi-dire* is able to wake up from the fantasy of ideology and recreate itself as an act. This is what Harold Bloom might call "strong theory," just as he distinguished poetry that accomplished similar ends "strong poetry."[20] Because optics and identity are the components, according to Freud, of the uncanny (*Unheimlich*), it is inevitable that theory – even non-architectural theory – revolve around the primary *architectural* theme of the home, and the manic melancholy of the scholar will forever be directed towards the project of distinguishing (or not) the temple from the labyrinth.

Notes

1. The well-known "Who's on first?" dialogue was devised by the comedy team Bud Abbott and Lou Costello. One of the many sources of this transcript is http://www.baseball-almanac.com/humor4.shtml (accessed 20 February 2009). "The general premise behind the exchange has Costello, a peanut vendor named Sebastion Dinwiddle, talking to Abbott who is Dexter Broadhurt, the manager of the mythical St. Louis Wolves. However, before Costello can get behind the plate, Abbott wants to make sure he knows everyone's name on the team." My use of this quote in the context of the famous mirror-stage idea of Jacques Lacan and the possible anticipations of this idea by the eighteenth-century Neapolitan philosopher of culture Giambattista Vico aims to show how the idea of time, generated in both a backward and forward direction by the mirror stage, creates a perplexity at both the level of the phenomenon and the science of that phenomenon. The idea of the subject ambiguously known through a proper name and pronoun echoes throughout history, the Homeric case of Odysseus's trick played on the Cyclops, giving his name as "Nobody," affords an escape from the giant's impossible prison. This is the first of many instances where the "defective subject" becomes both a key and a password.
2. Jane Gallop, "Lacan's 'Mirror Stage': Where to Begin?," *SubStance* 11, 4, Issue 37–38, A special issue from the Center for Twentieth Century Studies, 1983, pp. 118–128.
3. Irish novelist James Joyce famously used an imagined transcription of this thunder word in his novel, *Finnegans Wake*, New York: Viking Press, 1967, p. 3. "Ababadalgharaghtakamminarronnkonnbronntonnerronntuonnthunntrovarrhounawnskawntoohoohoordenenthurnuk!" There are nine other "thunder words" in the novel.
4. The alternative, that Vico actually "died," is not considered by any Vico scholar to my knowledge. This would engage the Lacanian idea of "between the two deaths," which was in Vico's age the popular device of the death narrative, the *katabasis*. The most famous source would have been Macrobius's *Commentary on the Dream of Scipio*, an elaborate discussion of Cicero's retelling of the dream of the nephew of Scipio Africanus, who was led by his dead uncle to the ramparts of Elysium to see the world below "as it really was." This involved the Platonic reversal of life and death. What for the body seems to be life is really, for the soul, death; and the body's death is the birth of the soul. Reading *The New Science* and Vico's *Autobiography* as death narratives solves many textual problems and empowers Vico with the magisterial powers of an Orpheus. These powers do not include omniscience but, rather, a combination of clairvoyance with idiocy, in keeping with Vico's self-diagnosis of his melancholic-choleric personality. Giambattista Vico, *The New Science of Giambattista Vico*, Thomas Goddard Bergin and Max Harold Fisch (trans.), Ithaca, NY: Cornell University Press, 1984; Giambattista Vico, *The Autobiography of Giambattista Vico*, Max Harold Fisch and Thomas Goddard Bergin (trans.), Ithaca, NY: Cornell University Press, 1944.
5. Mark Linder, in his review of Lorens Holm's article describing Brunelleschi's famous perspective mirror experiment, seems to me to make an inverted use of the multiple coincidences at the "ontic and ontological" levels, as Heidegger might put it. Mark Linder, "Time for Lacan: Looking after the Mirror Stage," *Assemblage* 21, August 1993, pp. 82–83; Lorens Holm, "Reading through the Mirror: Brunelleschi, Lacan, Le Corbusier: The Invention of Perspective and the Post-Freudian Eye/I," *Assemblage* 18, August 1992, pp. 20–39. Linder takes seriously Gallop's report of Lacan's warning to his students: "It happens that our students delude themselves in our writings into finding 'already there' that to which our teaching has since brought us"; Jacques Lacan, *Écrits*, Paris: Editions de Seuil, 1966, pp. 94–95. Catherine Clément, in her book *The Lives and Legends of Jacques Lacan*, New York: Columbia University Press, 1983, affirms as much. But, what Linder considers to be a case of simple anachronism is something more in Gallop's view: "Clément's claim that all Lacan is *en germe* in 'The Mirror Stage' produced an enthusiasm in me which immediately became embarrassing. My embarrassment corresponded to a realization that it was extremely pleasurable to find the later Lacan 'already there' in the early writing. An anticipation of maturation produced joy along with a willingness to suspend disbelief. This joy may resemble the 'jubilation' which Lacan ascribes to the child assuming his mirror image, being captivated by an analogy and suspending his disbelief" (Gallop, *op. cit.*, p. 120). In other words, this is no "mere

mistake" but a moment of synchronicity, unmistakable because of its parallel geometries, effects, and error. The student anticipates in the early texts the maturation of Lacan's teachings. Thus, somehow, the effect of Lacan's text on his students is analogous to the effect of the mirror on the infant. Lacan's text functions as a "fictional" but effective mirror image. Holm would be more in agreement than Linder with Gallop's point since Gallop does not condemn the bidirectionality of the mirror stage's temporality but, rather, connects it to the scholar's *necessary* complicity. Linder plays Zeuxis to Holm's Parrhasius by mistaking the "curtain": the ability of the mirror used by Brunelleschi to back-project a history of error and failure, for which the "event" of the perspective device promised a utopian correction.

6   This image may be one of the first philosophical deployments of the parlor game, "blind man's buff" (or "bluff"), since the blind Homer is credited with sublime insight because he is able to represent partiality (invisibility/monstrosity) directly. Homer's representational ability is thus 1:1, but in the special sense that liminality is transferred to poetry without distortion.

7   Ed Pluth, *Signifiers and Acts: Freedom in Lacan's Theory of the Subject*, SUNY series, Insinuations: Philosophy, Psychoanylsis, Literature, ed. Charles Shepherdson, Albany, NY: State University of New York Press, 2007, p. 36.

8   Jane Gallop, *op. cit.*, p. 121.

9   For a perceptive overview of the cultural phenomenon of emblem books, see Don Cameron Allen, *Mysteriously Meant, the Rediscovery of Pagan Symbolism and Allegorical Interpretation in the Renaissance*, Baltimore, MD: Johns Hopkins Press, 1970.

10   Donald Phillip Verene, *Knowledge of Things Human and Divine: Vico's* New Science *and* Finnegans Wake, New Haven, CT, and London: Yale University Press, 2003, p. 146.

11   I would like to go further into this, but the economy of this essay requires a compact explanation. By "voice" I mean one of Lacan's famous four "partial objects," objects that, as components of the "partial drive" (the drive whose goal lacks an aim and circles back to the same empty place vacated by the "impossible" object — the breast, the gaze, feces, the voice). In his excellent review, Mladen Dolar explains carefully that the voice defies phonemic analysis and, in popular experience as well as in psychoanalysis, radically delocalizes speech. Mladen Dolar, *A Voice and Nothing More*, Cambridge, MA, and London: MIT Press, 2006. Dolar extends the important earlier work of Michel Chion on the "acousmatic" voice in cinema. See *The Voice in Cinema*, Claudia Gorbman (ed. and trans.), New York: Columbia University Press, 1999. I realize that the authorial voice is only acousmatic or a partial object by analogy, but I am interested in the imaginary embodiment of the voice in the mind of the reader as a special category. It constitutes the "minimal element of ventriloquism" that Dolar finds to haunt every speech act.

12   Margherita Frankel, "The 'Dipintura' and the Structure of Vico's *New Science* as a Mirror of the World," in *Vico: Past and Present*, ed. Giorgio Tagliacozzo, Atlantic Highlands, NJ: Humanities Press, 1981, pp.43–51.

13   The trope of the melancholy genius was widely known in Vico's day, from humoristic lore that had established links connecting cosmological, medical, poetic, and behavioral qualities. Authors could draw on it to set their work and personalities within a preferred heroic stance. See Raymond Klibansky, Erwin Panofsky, and Fritz Saxl, *Saturn and Melancholy, Studies in the History of Natural Philosophy, Religion, and Art*, New York: Basic Books, 1964.

14   Some of this ground has been covered by Zakiya Hanafi, *The Monster in the Machine: Magic, Medicine, and the Marvelous in the Time of the Scientific Revolution*, Durham, NC, Duke University Press, 2000.

15   Ed Pluth, *Signifiers and Acts*, p. 63–64: "To understand something about desire in Lacan's work, we have to see how desire comes about from the experience of not having a place in the Other after one already experiences having a place. ... Instead of using the distinction between demand and desire to think about this difference, Lacan made a distinction between fantasy and act. ... [W]hat Lacan calls an act is something that cannot be reduced to a demand structure: in fact, the difference between fantasy and act [Vico: melancholy and choler] is defined ... by the presence or absence of demand. An act, unlike the fantasy, does not make a demand on the Other and is not aimed at acquiring recognition by the Other." It is precisely this same division that marks off the text-as-fantasy of *The New Science* from the frontispiece-as-act, or more generally Vico's *The New Science* from his *Autobiography*, where the former makes a minimal but impassioned

gesture to the reader and the latter boldly constructs a false self outside the demands of the Neapolitan intelligentsia that was his other: a case of Lacanian "false false," the negation of a negation that becomes a form of *mi-dire*, or half speech. Note also the Edenic undertones of Pluth's characterization of desire as coming about through the experience of "not having a place in the Other after one already experiences having a place."

16 It is interesting to compare Vico's swiddens, which have an anthropological basis, to the biblical story of the Garden of Eden, where God's presence is limited to the sound of rustling footsteps. In film theory, the "acousmatic" or off-screen voice is defined by the quality of its absence. In Eden, the issue of voice is related to blindness; Adam's invisibility follows directly after his violation of "God's word."

17 The publication of the new edition of *The New Science* was to be underwritten by the Venetian architectural theorist, Carlo Lodoli. Lodoli withdrew his support after a disagreement, leaving open the matter of what to do with the pages reserved for the dedication. Vico was dissuaded from publishing his contentious correspondence with Lodoli. Instead, he claims that the frontispiece and its explanatory materials were assembled at the last minute as a substitute. The well-thought-out composition of the image known as the *dipintura* and the tight composition of the commentary cast some doubt on this explanation. Although many elements of the Rosicrucian-style image were in popular circulation through the emblem books of the day, some devices (the perched globe, the cracked statue) were specific to Vico's theory. Also, the text's silence on the presence of the helmet of Hermes, which coordinates later textual references to the role of the reader, would have required some prior planning. The *dipintura*'s general role is that of the subject in front of the mirror, pulling itself erect, able to see through the scattered symbols towards a unified idea of culture in the text beyond but aware of its fragmented nature through the hypotaxis of accidental objects in the field of the image.

18 Lacan was careful to point out that transitivity was an important indicator of the mirror stage's transformation of space's causal function. After the mirror stage, children frequently commit an error of transitivity. Hitting another child, the perpetrator will say "He hit me!" That this account is an error in need of correcting in order to understand the proper order of agents, causes, and effects shows how the mirror stage's primary error is just such a confusion. The effect of unity/mastery is presented prematurely. The condition of fragmentation is back-projected through the mechanism of the future anterior tense. Inside and outside, cause and effect, past and future are a part of the symbolic order, but the subject is split by the bar that places it in both the symbolic and the imaginary. Behind the screen of the subject as "photo-graphed" is the subject-in-pieces. Add to the child's misconception of causality the transitivity (error) of blindness/invisibility. A child who puts a bag over his/her head imagines him/herself to be invisible rather than blind. Blindness can be indicated also by the invisibility of the head, a common motif in eighteenth-century emblems showing Justice or other allegorical figures as headless, when in truth their head has been made invisible from the point of view of those below, who cannot themselves see the invisible realm penetrated by the head of the figure.

19 A full program of Hermes' interests is given in Norman O. Brown's excellent study, *Hermes the Thief, the Evolution of a Myth*, Madison, WI: University of Wisconsin Press, 1947. Brown notes that Hermes' highly varied qualities and powers derive from the central function of the boundary, particularly the boundary between life and death.

20 Harold Bloom, *The Anxiety of Influence*, New York: Oxford University Press, 1973.

Part 4

Part 4

The challenges of instrumental knowledge

In the last two centuries the long tradition of humanist culture has been seriously challenged by the apparent efficiency and adequacy of instrumental representations and models. The chapters presented in this part highlight how architecture and design provide powerful vehicles to redefine the role of the humanities in the modern age, against a background of a tendency towards the internationalisation of culture and the instrumentalisation of life, politics and power. Each study looks in different ways at how architectural representation opens up the possibilities of a sustained and meaningful dialogue between a "past" – variously defined by mytho-historic or paradigmatic references – and the immediate conditions of modern or contemporary life.

In the opening chapter, "Architecture as a humanistic discipline", Dalibor Vesely challenges our assumptions of place and space, by suggesting there must be continuity between philosophical understanding and history. A thorough and ongoing interpretation of our lived existence is required as a means of deepening our understanding of the historical past. Where we are affects how we feel; to be "in" a mood, for example, implies a spatial context, an environment and therefore a situation. As Vesely describes, in the journey from cosmology to astronomy we discover the "emptiness" of space but are also captivated by its mystery, as conveyed for example in the work of the Surrealists. Nicholas of Cusa suggests that the more we are aware of the gaps in our knowledge, the more we can acquire learned ignorance. The examples given in this chapter reveal that, although it may be possible to list the items, pieces of furniture, quality of light and scenery in the particular situation such as a French café, none of these descriptions can adequately express the experience of actually being there; nor can they convey the anticipation or imagining of that experience. By emphasising the situational nature of architecture Vesely makes a case for its role as a "humanistic discipline".

In the second chapter, "The situational space of André Breton's atelier and personal museum", an artist's studio is taken as an example of both a personal museum, and the process of generating new forms of thought, in an exploration of the collection of the Surrealist artist André Breton. Dagmar Weston suggests that it is impossible to "recreate" a studio within a museum, and that individual objects contribute less than the whole in situ. In Breton's studio the objects of the everyday merge with medieval reliquaries, thereby creating a sense of aura to Weston's description of Breton's atelier, which highlights the life-world as a temporal dimension of situated spatiality and questions whether it is possible to assemble, as an isolated art object, a place that is the embodiment of the creative arts.

In "*L'Histoire assassinée*: Manfredo Tafuri and the present", Teresa Stoppani explores Tafuri and the "history of the present". As history reverses itself to the present, the "past" becomes a project and the "contemporary" conflates with "modernity". As Stoppani states: "'Contemporary' is a term that Tafuri would most likely refuse, as for him the 'contemporary' is still part of a long and far from resolved modernity. Tafuri would rather speak of the 'present'". He challenges moreover the suggestion that change is embedded in history, and his "historical project" is in fact an open mapping of the production, recurrence and re-emergence of principles or ideas. Stoppani defines Tafuri's "project" as a history of architecture that is independent from both architectural practice and the history of art. Called "operative history", the author explains that Tafuri saw criticism not as a "tool" that can be applied to historical interpretation but rather is an integral part of history.

In the next chapter, "Nature choreographed", Anne-Louise Sommer offers a fresh approach to the often overwrought discussions of aesthetics. Her description of Pavlovsk, a palace and garden built by Catherine the Great near St Petersburg for her son Pavel, the future czar, considers both principles of eighteenth-century garden design and the processes by which it was conceived, created and experienced. While we might assume that there is nothing more authentic than nature, Sommer suggests that authenticity is about our experience of a place, rather than the creation or recreation of buildings or landscapes as "set pieces"; where built form anticipates the ruin in much the same way that the "natural" is designed, constructed, destroyed and recreated. Human intervention that relies on artifices such as theatrical-scene design – through false perspective or *trompe l'oeil* painting – retains in the landscape garden a "natural" state that is being endlessly transformed.

In the final chapter, "The birth of modernity out of the spirit of music? Henry van de Velde and the Nietzsche Archive", Ole Fischer examines conditions and qualities of design as an expression of philosophical thought, in particular that of the German philosopher, Friedrich Nietzsche. Towards the end of his life, an archive was created to display his works, and, while he was alive, the philosopher himself. Fischer looks critically and speculatively at how the Nietzsche Archive thematises the translational process from philosophical critique to architectural design. The architect in question is Henry van de Velde who, commissioned by Elizabeth Förster-Nietzsche, the philosopher's sister, designed the Nietzsche Archive that was a home and part-time shelter for the philosopher. The archive was created in the "new style", in response to conflicts of the late nineteenth century. Van de Velde's design approach coincided with Förster-Nietzsche's call for the "New Weimar" to leave behind its past, as represented by Goethe, and to give birth to a new age of arts, architecture and life, emblematic of Nietzsche's concepts of modernity.

Chapter 16

Architecture as a humanistic discipline

Dalibor Vesely

Introduction

Today architecture is treated most often as a technical discipline dominated by the knowledge of science, seen as the ultimate criterion of truth and what is real. We are familiar with the reference to the "real" world that we hear, almost as a cliché, so often today. What is meant by *real* is usually the pragmatic reality of the everyday, defined by current politics, economics, market forces and technological developments. On a more sophisticated level, the real is defined by the criteria of science. It is a standard and repeated experience that, in case of doubt, we ask science to be the ultimate arbiter of truth.

And yet there is still another sense of what is *real* that we encounter in our everyday life, in personal relations, in friendships, in our judgements of what is good or bad, in our faith etc. It is mostly our intuition and common sense that point to a different reality which we share to a great extent, but consider, nevertheless, mostly as personal or subjective and find it therefore difficult to give it an objective validity.

It is partly for that reason that some contemporary thinkers speak about modern citizens as citizens of two worlds. This is related only indirectly to the well-known discussion of C.P. Snow, outlined in his influential book *The Two Cultures*.[1]

Hans-Georg Gadamer has this to say about the citizens of two worlds:

> The forming of European civilisation by science implies not only a distinction, but brings with it a profound tension into the modern world. On the one hand, the tradition of our culture, which formed us, determines our self-understanding by means of its linguistic–conceptual structure which originated in the Greek dialectics and metaphysics. On the other hand, the modern empirical sciences have transformed our world and our whole understanding of the world. The two stand side by side.[2]

Ambiguity of the two worlds

The two worlds stand side by side in principle, but in everyday life they very often overlap. The result is a state of ambiguity. This can be best illustrated in our own field by concrete examples. In the work of Daniel Libeskind, the critical, analytical vision very often overlaps with the romantic and personal inspiration. In his own words:

> Contemporary formal systems present themselves as riddles – unknown instruments for which usage is yet to be found. Today we seldom start with particular conditions which we raise to a general view; rather we descend from a general system to a particular problem.[3]

This is complemented by the following statement: "[E]ver since I encountered Johannes Kepler's study of the six cornered snowflake I have marvelled at the infinity of crystals that nature created in every snowflake. Let me leave you with this thought: All architecture is crystalline."[4]

Perhaps even more explicit illustrations of the state of ambiguity in our field can be found in the work of Leonid Leonidov. Leonidov's first period, defined by the transparent constructivist rationality, differs fundamentally from the second, defined by personal experience, cultural memory and different levels of freedom to bring them to visibility. It is not easy to describe clearly the nature of the difference, particularly in view of the second period. There is, no doubt, an overlap of certain elements, but the transparent engineering approach melts into a dreamlike vision motivated by personalised human and cosmic motives and values.

The state of ambiguity changes once we move from the everyday to the more institutional levels, particularly to the domains of academia and explicit knowledge. If we leave behind the tendency to subordinate humanities to the methods of the natural sciences, we discover that the division between science and the humanities

16.1
Daniel Libeskind.
Extension of Victoria
and Albert Museum,
London

16.2
Ivan Leonidov. Headquarters of Heavy Industry, Moscow (1934) and City of the Sun (1943–1957)

is taken for granted as a fait accompli. This division has its origin in the nineteenth century when the dilemma of science and the humanities was for the first time created. The relatively unified culture was at that time challenged by a radically new knowledge based on the mathematical criteria of truth. How did it come about?

Mathesis universalis

The new knowledge has its origins in the reaction to the sixteenth-century religious and cultural relativism and in the search for the new universal foundations of faith and truth. Such foundations were found in the universality of mathematics (*mathesis universalis*), referred to at that time very often as the "queen of the sciences" (*regina scientiarum*), elevated sometimes in such names as *ars magna, ars divina* or *scientia divina*.[5] These lofty names could not convince without supporting evidence from the physical world in which it became possible to speak of physics and metaphysics in the same terms, and theological problems could be, by implication, treated as metaphysical and eventually as physical (*theologia naturalis*).[6]

Universal mathematics (*mathesis universalis*) did lay claim to cover the same area of knowledge as traditional logic – in other words, the area of all possible knowledge. The essence of the new knowledge was made, and seen as, compatible with the language of mathematics. The new knowledge was, in the end, turned into a universally applicable paradigm based on Newtonian mechanics. The consequences were summarised by one modern thinker in the following way:

There is something for which Newton – or better to say not Newton alone, but modern science in general – can still be made responsible: it is the splitting of our world in two …. It did this by substituting for our world of quality and sense perception the world in which we live and love and die, another world – the world of quantity, of reified geometry, a world in which though there is a place for everything, there is no place for man. This, the world of science – the real world – became estranged and utterly divorced from the world of life which science has been unable to explain – not even to explain away by calling it subjective …. True, these worlds are every day more and more conceited by the praxis. Yet for theory they are divided by an abyss. Two worlds, this means two truths. Or no truth at all. This is a tragedy of modern life which solved the riddle of the universe but only to replace it by another riddle, the riddle of itself.[7]

The dominating role and monologue of the Newtonian paradigm left the humanities in a deep shadow. This can be best illustrated by the new seventeenth-century attitude to poetry – the paradigm and essence of the humanities. Poetry was, at that time, considered to be rather dead, as the following statement illustrates: "Prose is quite able to express anything you can say in verse; it is more precise, more to the point and it takes less time."[8]

And another contemporary opinion: "When you begin reading a piece of poetry, remember you are reading the work of a purveyor of lies, whose aim it is to

16.3
Klementinum, Prague main library and Jacques Lajoue, *Le cabinet physique* (1752)

feed us on chimaeras or on truth so twisted and distorted that we are hard put to it to disentangle fact from fiction."[9]

However, despite their secondary role in culture – based on the ideals of the Enlightenment – the humanities sustained their own life, though rather fragmented and mosaic-like. In the whole pleiades of names, humanistic tendencies and movements, we should mention the works of Giambattista Vico, Friedrich Schleiermacher, the Romantics, the German historical school and W. Dilthey. All contributed to the formation of modern humanities, and in particular to the establishment of modern hermeneutics embodied in the writings of M. Heidegger and H.G. Gadamer.[10]

Giambattista Vico and common sense

Gadamer sees the primary role of humanistic tradition in *Bildung*, common sense, taste, and judgement (good sense). In the field of architecture the most relevant mediating link is common sense, which links the modern hermeneutics with the work of thinkers like Vico. As Gadamer states: "I have rightfully claimed for my own work the testimony of Vico."[11] And further in *Truth and Method*, Gadamer reminds us:

> There is something immediately evident about grounding philosophical and historical studies and the ways the human sciences work on the concept of the *sensus communis*. For their object, the moral and historical existence of humanity, as it takes shape in our words and deeds, is itself decisively determined by the *sensus communis*.[12]

Common sense is knowledge of the concrete and it is concrete knowledge because it is a sense acquired by living in a concrete community and determined by upholding the value of communal traditions. Common sense is historical in that it preserves tradition and not just as a datum of knowledge but as a principle of action. Common sense is closely linked with the meaning of commonplace and thus with the topology of being.

Vico writes: "Human choice, by its nature most uncertain, is made certain and determined by the common sense of men with respect to human needs or utilities, which are the two sources of the natural law of the gentiles."[13] Moreover, "common sense is judgment without reflection, shared by an entire class, an entire people, an entire nation, or the entire human race".[14] Vico's common sense can be seen as an anticipation of Heidegger's pre-understanding, and in the framework of our own argument; it is an anticipation of the structure of the latent world, manifested in typical situations.

The separation of science and the humanities

In science an attempt to understand the given phenomena observed in nature became a problem in the second half of the nineteenth century when the gap between the world of science and the natural world was for the first time strongly felt. It is interesting, but perhaps not entirely surprising, that the first to raise the issue of the gap were the scientists. One of them was Hermann Helmholtz, who, in a lecture delivered in 1862 and entitled "On the Relation of the Natural Sciences to the Whole of Science", drew a distinction between the natural sciences and the humanities, based not on different methods but on the difference of their subject.[15]

Another step, much closer to the formation of modern phenomenology, was the contribution of Richard Avenarius, who represents the culmination of nineteenth-century positivism in a form that he described as "empirio-criticism".[16] He was the first to use the term *Weltbegriff* (*der menschliche Weltbegriff*) to convey the human concept of world adopted by Edmund Husserl as *Lebenswelt* (lived world). What is interesting and rather profound is that Avenarius thought that all the different positions in our highly differentiated, syncretic world are only variations of the common, given natural world.[17]

I have mentioned the role of science in the formation of the specificity of the humanities to correct the prevailing view that phenomenology as a philosophy was at the origins of that formation. This became true for phenomenology only in its later stages. We should remember that Husserl himself was a mathematician before he became a philosopher. The first intention of phenomenology was to lay foundations for a rigorous, universal science, capable to resolve the crisis in the sphere of knowledge, manifested first as a problem of the foundations of modern mathematics and logic, leading only later to the problem of the new foundations of knowledge in general. The crisis of the foundation of science manifests itself also as an unresolved relation of the natural sciences and the humanities, and, later, as a question of truth in simulations, in virtual realities and artificial intelligence. It is perhaps not surprising that the first serious attempt to challenge the methodical monologue of positive empirical sciences came from the humanities and in particular from history in its effort to overcome the dogma of historicism. The best examples of such an attempt are the written works of Gottfried Semper and the late work of Wilhelm Dilthey, who happened to appreciate Semper's *Der Stil*, which he describes as an "enlightening model for how an important historical problem could be solved in aesthetics" and considers Semper for that reason to be a real successor to Goethe.[18] There are many similarities between Semper's and Dilthey's thinking, such as the search for the simplest constituent elements of history and the use of comparative method, but most of all the search for the scientific theory of history and art. For Dilthey the paradigm of method and its scientific legitimacy was Kant's *Critique of Pure Reason*, which he tried to emulate with his own *Critique of Historical Reason*.[19]

However, the result was not a success. Dilthey's articulation of the historical reason did not help to overcome the conflict between a recognition of the historicity of the human world and the attempt to develop a method which could grasp the phenomenon of historicity. As Gadamer demonstrated, the second ambition cancels the first. Dilthey's contribution, and in some sense Semper's as well, represents a turning point in the theoretical thinking about arts and the humanities.

Their search for the scientific method in these fields led to the demonstration of the limits of such enterprise and opened the door for a very different way of thinking about the humanities and arts, including architecture.

The new way of thinking was opened and cultivated mostly in the sphere of phenomenology and, later, in hermeneutics. In this traditional method, epistemology and theory are not entirely dismissed but became rather redundant. Their place is taken over by a more critical and subtle thinking, sensitive to the hermeneutical conditions of typical situations, reciprocity of their universal and particular meaning, formation of historical concepts and their creative interpretation. In the last years of his life Dilthey came very close to Husserl and phenomenology. Though it was rather too late, his influence on the phenomenological movement, particularly on Heidegger and Gadamer, was substantial.

The sequence of Dilthey's influences illustrates the changes in the nature of phenomenological thinking and its development from the immanent transcendentalism of Husserl, to Heidegger's and Gadamer's ontology based on language and to the most recent stage of thinking, situated in the sphere of embodiment (corporeality) articulated most consistently so far by Jan Patocka, Maurice Merleau-Ponty, and his followers Jean-Luc Marion, Barbaras and others.

Contemporary foundations of humanities

The critical terms of the phenomenological foundations of the humanities are the typicality of experience, based on praxis and embodied in typical (paradigmatic) situations. The "typicality of experience" is a sedimented experience which always precedes a particular design decision and its form. Reading, for instance, is essential to the vision of a library, but always transcends it. It can take place elsewhere or even without it. On the other hand, the concept of a library without a vision and understanding of the conditions of reading, borrowing of books and the inner life of the library is empty. Library seen as a type is not an original reference; it is always preceded by the typicality of particular experience of using the library.

The typical experience has its origins in praxis. The classical notion of praxis belongs to the fundamental constitution of human beings and their situation in the world. Praxis does not depend on the abstract knowledge of norms but is always concretely (that is, practically) motivated.

> In every culture a series of things is taken for granted and lies fully beyond the explicit consciousness of anyone, and even in the greatest dissolution of traditional forms, mores and customs, the degree to which things held in common still determine everyone is only more concealed.[20]

The deeper meaning of praxis is living and acting in accordance with ethical principles. For a more specific understanding, it is necessary to see praxis as related to a particular place, where people are not only doing or experiencing something, but which also includes things that contribute to the fulfilment of human life. This includes everything associated with human activity, for instance the table on which we take our daily meal or the walls of a room which protect the intimacy of our conversation.

The relation of praxis to a particular place leads to a formation of *typical (paradigmatic) situations*. Situations act as receptacles of experience and of those events which sediment a meaning in them, a meaning not just as survivals and residues, but as the invitation to a sequence of future experiences. Situations endow experience with durability in relation to which other experiences can acquire meaning and can form a memory and history.

Situations are dependent and closely related to habits, traditions and customs. Their nature is similar to the nature of institutions, deep structures and archetypes. Situations represent the most complete way of understanding the condition of our experience of the surrounding world and the human qualities of the world. The close link between praxis and typical situations indicates that praxis always belongs to a world it articulates and thus brings about.

This level of articulation can be reached in a hermeneutical interpretation of the lived world as it is manifested in typical situations. On that level articulation can be read as a language structured by similarities, metaphors and analogies, which may be described altogether as poetic paradigm. Poetic paradigm is a key to the understanding of the lived (natural) world in its wholeness. The natural world is not a thing or a sum total of things that can be seen or studied in their explicit presence as isolated entities. It is an articulated continuum to which we all belong. The main characteristic of the natural world is its continuity in time and space and its permanent presence. A most explicit manifestation of this can be seen in language, revealing most clearly the structure of the natural world. In this world, we can move, in the same way as in language, into the past or future, survey different regions of reality, refer to almost anything in our experience and translate the experience into any language. This possibility includes also the language of science and religion. What is revealed in language points to a level of articulation, which shows our involvement in the structuring of the natural world as well as the deepest level of reality, where language meets the natural conditions, that is, the given reality of embodiment in its most elementary form.

The role of the poetic paradigm in understanding the natural world in its wholeness is recognised not only by humanists and artists, but also by scientists. Werner Heisenberg refers to it when he speaks of the "one" (common good), which is only a different name for the unifying role of praxis. He writes:

> In the last resort, even science must rely upon ordinary language, since it is the only language in which we can be sure of really grasping the phenomena ... the language of images and likenesses is probably the only way of approaching the "common good" from more general domains. If the harmony in a society rests on a common interpretation of the "common good", the unitary principle behind the phenomena, then the language of poetry may be more important here than the language of science.[21]

16.4
David Thompson. Paris project, Film Resource Centre (2005)

Architecture as a humanistic discipline

From all that has been said so far, we can draw already a preliminary conclusion, that architecture is not in the first place a technical but a humanistic discipline. This must be clear to everyone who sees a distinction between means and goals, and agrees that the goal, the essence of architecture, its main purpose, is to situate our life in a particular place and create the right conditions for our existence and coexistence, not only with other people, but also with the given natural conditions and cultural circumstances. Skills, techniques and technologies are only means that can help us to fulfil this purpose and goal.

Unlike individual techniques and specialised sciences, architecture is faced with reality in its full, given phenomenal presence. In any kind of design we have to address a particular space in its totality. We can concentrate on a particular aspect, light, material, events etc., but in the end we can judge and appreciate the results only in the context of the space in its unity and as a whole. The unity of space does not depend on the integrity of individual spatial aspects only, but also on their reciprocity, how changing light influences the materiality of the space for instance. The possibility of grasping the space in its wholeness was, and to a great extent still is, a privilege of our imaginative abilities to hold together a coherent vision of space. However, under the pressure when experience is supposed to be presented in a form of explicit knowledge, an unanswered question opens: Where is the discipline that can grasp the full reality of space and elevate it to an explicit level of understanding? It is quite obvious that we don't have a science that can claim such ability. This is true not only for the specialised sciences but also for the humanities as they are on average practised today. The most likely discipline that can help us in the task of a better understanding of space is, as we have already seen, philosophy, particularly in its phenomenological and hermeneutical orientation.[22] This includes the possibility to use partial results of individual sciences and humanities subjected to a critical and creative interpretation.

To the results of science and the humanities we should add the contributions made by other areas of culture, such as painting, sculpture, literature, theatre etc. Architecture is not only influenced by different areas of culture but it is also created there, though only indirectly. The mediated, indirect contributions to the making of architecture increase the possibility that some aspects of architectural reality (space) are more clearly visible in the non-architectural areas of culture. Because a higher and more explicit level of articulation, painting, sculpture, theatre and literature can tell us about the role of light, nature and materiality of space and about the role of particular events in the formation of space more than isolated architectural studies on their own.

The final results and success of design depend on bridging the distance between the contributions of individual disciplines and the unifying understanding. The unifying understanding, we may already conclude, is an achievement of a humanistic approach; what makes this approach to design possible is not only the unifying nature of knowledge, but also the paradigmatic nature of typical human situations.

Paradigmatic nature of typical situations

The term "paradigmatic" refers to the power of typical situations to bring and hold together a vast richness of human experiences and give them relative stability. The stability of situations is revealed in habits, customs and traditions. It is a source of constant surprise to see to what extent our life is structured by the typicality of situations of our everyday life, such as eating, work, learning etc., situated in typical places.

The process that constitutes and preserves the typicality of situations can be described as a continuity of reference to the ultimate source of stability in the given natural (cosmic) conditions of our world and its history.[23] However if we want to appreciate the real nature of situations we must turn to a concrete example.

Figure 16.5
A French café

If we look closely at a French café for instance, it is obvious that its essential nature is only partly revealed in its visible appearance. For the most part, it is hidden in the field of references to the social and cultural life related to a particular place.

Any attempt to understand the character, identity or meaning of the situation and its spatial setting using conventional typologies is futile. The essential reality of the situation is not entirely revealed in its visible appearance. It cannot be observed or studied just on that level.

Its representational structure can be grasped through a pre-understanding based on our familiarity with the situation and with the segment of world to which it belongs. Pre-understanding in this case is a sedimented experience of the world, acquired through our involvement in the events of the everyday life. The identity of the French café is to a great extent defined by its institutional nature, rooted in the habits, customs and ritual aspects of French life. The formation of identity is a result of a long process in which the invisible aspects of culture and the way of life are embodied in the visible fabric of the café in a similar way as is language in the written text. The visible "text" of the café reveals certain common, deep characteristics, such as its location, relation to the life of the street, transparency of enclosure and a certain degree of theatricality expressed in the need to see the life of the outside world, but also a need to be seen in it like an actor, the ambiguity of inside and outside expressed not only in the transparency of enclosure, but also in the choice of furniture etc. These are only some of the characteristics that contribute to the identity and meaning of the French café as a culturally distinct typical situation.

The task to grasp the essential nature and typicality of situations is critical not only for the making of architecture, but also for its interpretation in reception and use. This brings us to a final conclusion that we don't need to speak about the influence of the humanities on architecture or about a separate role of the humanities in design, because architecture itself is a humanistic discipline and should therefore be treated as such.

Notes

1   C.P. Snow, *The Two Cultures*, Cambridge, Cambridge University Press, 1993.
2   Hans-Georg Gadamer, *On Education, Poetry and History, Applied Hermeneutics*, D. Misgeld and G. Nicholson (eds.), L. Schmidt and M. Reuss (trans.), Albany, NY: State University of New York Press, 1992, pp. 209–21.
3   Daniel Libeskind, *End Space*, London: Architectural Association, exhibition catalogue, 1980, p. 20.
4   Daniel Libeskind, *Breaking Ground, The Journey from Poland to Ground Zero*, New York: Riverhead Trade, 2005, p. 230.
5   Nicholas Jardine, "Epistemology of the Sciences", *The Cambridge History of Renaissance Philosophy*, Cambridge: Cambridge University Press, 1988, pp. 694–96; Giovanni Crapulli, *Mathesis universalis, geneesi di una idea nell XVI secolo*, Rome: Edizioni dell'Ateneo, 1969.
6   Charles H. Lohr, "Metaphysics," *The Cambridge History of Renaissance Philosophy*, Cambridge: Cambridge University Press, 1988, pp. 616–20.
7   Alexander Koyre, *Newtonian Studies*, London: Chapman & Hall, 1965, p. 24.
8   Houdar de la Motte's commentary on Virgil's Ode, quoted in Paul Hazard, *The European Mind 1680–1715*, London: Penguin Books, 1953, p. 386.
9   *Ibid.*, p. 387.
10 Richard E. Palmer, *Hermeneutics, Interpretation Theory in Schleiermacher, Dilthey, Heidehher and Gadamer*, Evanston, IL: Northwestern University Press, 969.
11 Gadamer, "Reply to Donald Phillip Verene", *The Philosophy of Hans-Georg Gadamer*, The Library of Living Philosophers, vol. XXIV, Chicago, IL: Carus Publishing Company, 1997, p. 154.
12 Gadamer, *Truth and Method*, London: Shed and Ward, 1979, reprint 1985, p. 22.
13 *The New Science of Giambattista Vico*, trans. of the 3rd edn (1744), T.G. Bergin and M.H. Fisch (trans.), Ithaca, NJ: Cornell University Press, 1961 (1988), XI. (141) p. 63.
14 *Ibid.*, XII. (142), p. 63.
15 Gadamer, 1992, p. 40.
16 Richard Avenarius' (1843–1896) empirio-criticism, a radical version of late empiricism, which can be also seen as a refined version of positivism, came close to a discovery of the phenomenal reality of the world as a common ground for all different interpretations of the world itself.
17 Avenarius, *Der Menschliche Weltbegriff*, Leipzig: O.R.Reisland, 1891.
18 Wilhelm Dilthey, *Poetry and Experience*, R.A. Makkreel and F. Rodi (eds.), Selected Works vol.V., New Haven, NJ: Princeton University Press, 1985, p. 204.
19 Gadamer, 1979, pp 206–14.
20 Gadamer, *Reason in the Age of Science*, F.G. Lawrence (trans.), Cambridge MA: MIT Press, 1981, p. 82
21 Werner Heisenberg, *Across the Frontiers*, Peter Heath (trans.), New York: Harper and Row, 1974, pp. 120–21.
22 Gadamer, "The Universality of the Hermeneutical Problem," *Philosophical Hermeneutics*, D.E. Linge (trans.), Berkeley, CA: University of California Press, 1976, pp. 3–18.
23 Jan Patocka, *Body, Community, Language, World*, James Dodd (ed.), E. Kohak (trans.), Chicago and La Salle, IL: Open Court, 1998, pp. 143–53.

Chapter 17

The situational space of André Breton's atelier and personal museum in Paris

Dagmar Motycka Weston

Introduction

Setting out to propose a richer philosophical alternative to the reductivist and instrumental thought of nineteenth-century positivism, Surrealism (as the heir to the Romantic legacy) emphasized the key role of the imagination and personal creativity. Its own explorations were immersed in a rich background of the humanities, with philosophy, drama and literature chief among them. The sought-after reconciliation of "dream" and "reality" was to be achieved through reawakening a more primary sensitivity to the poetic dimension underlying concrete daily experience. In its interest in the phenomena of perception and its aim of describing the textures of primary human experience, Surrealism manifested many affinities with phenomenology. The central theme of Edmund Husserl's later writings is the need to overcome the conceptual bias of Western thought. Human being is to be understood as rooted in the life-world (*Lebenswelt*), the world of daily life, or reality as it is directly and immediately given in human experience. In contrast to the abstracted, scientific understanding of the universe, the life-world is immediately familiar to us and concrete. Phenomenology arose as an attempt to provide a unified foundation to the splintering views of reality being developed by the different branches of science, and to emphasise the importance of primordial perception and embodiment in human experience as a ground for understanding. The Surrealist affinities with phenomenological philosophy are manifest in the movement's concern with such "primitive" phenomena as the experience of space and time, memory, play, dance, humour and the erotic. Both also rejected abstract mathematical space as

17.1
André Breton at his writing table,
photo by Henri Cartier-Bresson,
c. 1960

incompatible with human experience of the world, describing instead a spatiality arising out of the conditions of embodiment. The Surrealists evolved what might be termed a "situational spatiality" – structuring spatial settings through the grouping together of evocative fragments.[1] This technique, which evolved primarily out of cubist and Dada collage, exploited the power of daily things to embody associations, memories and meanings, and thus to articulate a "world", a rich network of thematic relationships. André Breton created such a situational spatiality as a counterpart of his life and work.[2]

With many Surrealist artists this emphasis on personal creativity as part of everyday life took the form of configuring the domestic interior as a kind of mapping of the inhabitant's creative imagination. André Breton's studio is interesting not only as a fascinating and now tragically lost place. It is also a prime example of the personal museum and its reciprocity with an artist's life and work: an architectural embodiment of Breton's image of "communicating vessels". It manifests the central significance of collage and assemblage in structuring a "situational space", particularly with the primary role of spatial settings in the articulation of meaning. The Surrealist interpretation of these themes can fruitfully inform contemporary architectural discourse.

The dramatic social, political and cultural changes in the European industrial metropolis of the nineteenth century – the surging crowds, bold new technologies, accelerating consumerism and so forth – transformed the inhabitants' daily experience. At the same time, the architecture of the public realm was increasingly being drained of its symbolic and ethical dimension as the theatre of community life. One of the reactions to the new conditions of urban life was the shifting balance of public life and a growing sense of inwardness, with the private, bourgeois interior acquiring a new significance. As cultural life gradually became more internalized, the activities of interior design and collecting came increasingly to be seen as the appropriate vehicle to represent the owner's psychic landscape. This is evident in the house collections of a number of Parisian literary figures, including Honoré de Balzac, Victor Hugo and the Goncourt brothers. Inwardness is taken to an extreme with the neurotic interiority of Des Esseintes, the central character of J.K. Huysmans' 1884 novel *A rebours*, a work well known to Breton.

The preoccupation with psychic and perceptual experience, and the rise and popularization of neuropsychiatry in Paris in the last quarter of the nineteenth century led to attempts to project oneiric, pre-rational psychic states onto the bourgeois domestic interior.[3] This condition also informed Sigmund Freud's personal collection of antiquities, which he created in his Vienna study and consulting room in the early decades of the twentieth century.[4] Freud's assemblage of both tangible objects and immaterial things (like dream texts, jokes and slips of the tongue) constituted for him a kind of repository of fragments which became the raw material for his creative synthesis: his enquiry into the unconscious. With an intuitive understanding of the spatial dimension of thought and memory, Freud saw his personal museum as a kind of spatial embodiment of the human soul. Each one of his objects became for him a mnemonic device for a theme, making tangible the elements of thought.[5] At the same time, Freud's collection – occupying two rooms of an inward-looking Vienna apartment building – was a strong expression of a perceived need to create a complete, private culture in an increasingly barren world of the city. This famous collection was well known to Breton. Continuing these trends, collecting was very popular with many artists of the early twentieth century. Among Surrealist collectors were Paul Eluard, Max Ernst, Man Ray and Louis Aragon. The movement raised the process of finding, acquiring and grouping things together according to the principles of analogy to an essential part of their creative endeavour.

Modern culture's preoccupation with "primitivism" in art as a way of counteracting the perceived spiritual bankruptcy of official Western art, and of participating in a vital sacred tradition, is well known. The Surrealists were among the most avid admirers and collectors of such tribal and ritual artefacts, with a preference for the work of the Innuit, North American Indians and especially of Oceanic tribes.[6] In his collecting of primitive masks and fetishes, Breton emphasized their magical, symbolic and initiatory powers. One of the salient features of such artefacts was that, unlike Western high art, they were often very heterogeneous or *collagiste* in their make-up, with cheap, natural or perishable materials being assembled in ways which accentuated their metaphoric qualities. For instance, many of the masks in Breton's collection combined painted wood or clay with materials such as grass, animal hair, shells, bone, feathers, metal and beads. The use of such disparate materials to represent body parts (moss for hair, seashells for eyes) highlighted the primary analogical relationships between apparently distant things. Furthermore, the act of transplanting such "primitive" art into metropolitan galleries and artists' studios in itself generated the *frisson* of displacement, which was to be the essential ingredient of a Surrealist sensibility.

A related aspect of the "primitive" which informed cultural debates at this time was the thought manifested by "savage" or native cultures.[7] Theirs seemed to be an understanding of the human experience of the world which was more primary, primordial and concrete than was the case with the conceptual thought of Western positivist rationalism. This mode of thought, which Claude Lévi-Strauss called "the science of the concrete", sees the world in terms of magical powers and the sacred, and has a bias towards the specific, tending to classify things on the basis of analogy.[8] It has a relevance to the sensibility and spatial ordering of Surrealist collections.

Given the status since the Romantic period of the creative artist as a kind of secular saviour of a troubled culture, the *topos* of the artist meditating on the margins of society and preparing for its redemption through his creativity, became a pervasive one.[9] The mysterious mechanisms of original creativity, which springs spontaneously from the artistic imagination, became a matter of wide fascination. The artist is sometimes seen as an alienated, Bohemian figure, Christlike in his suffering. With the heightened interest in the depiction of daily life, and a fascination with the creative process which gives birth to art, the artist's studio as a theatre of life (and sometimes of philosophical speculation) became a particularly significant subject for painters during the second half of the nineteenth century. For certain twentieth-century artists, notably Pablo Picasso, George Braque and Henri Matisse, the studio setting as the locus of introspection and creativity became a major, life-long theme.[10] At the same time, there arose an intensification of interest in the mechanisms of the creative process, with the artist's preliminary work or personal possessions, for example, seen as valuable clues, and carefully preserved (often by the artist himself) in some form of archive or museum for future study and display. A key example of this phenomenon was the atelier-museum created by the symbolist painter Gustave Moreau in his house in the rue La Rochefoucauld in Paris in the 1890s. Introduced to this museum by his reading of Huysmans, André Breton (whose own apartment was a few blocks away) returned to it often during his life as a place of wonder, mystery and inspiration.[11]

Thanks in part to the spread of photography in the early decades of the twentieth century, the photo-portrait of the solitary artist in his studio – in his work clothes, immersed in thought or in the making of things, surrounded by his collection of objects and by the fertile clutter of the stuff of his creativity – became a defining image.[12]

In these noble pictures, the studio space itself – often rather hard-up and cluttered with the materials and products of imaginative life – embodies a liberated, Bohemian lifestyle, unfettered by bourgeois convention. It is also raised to a status approaching that of a secular temple, the matrix of godlike creativity, with the primitive fetishes looking on like terrible sentinels or domestic deities. The

Figure 17.2
Le Corbusier in his apartment at 20 rue Jacob in Paris, 1920s. Photograph by Brassaï.

artist's studio and home is also understood in these images as revealing his inner self.

This raising of the status of the creative artist (coupled with a newly found dignity of ordinary daily life) often implied the utopian corollary that *everyone* should now be able to engage in creative action. Breton, for example, gave prominent place to Lautréamont's dictum that everyone should create poetry. This demand was ideally satisfied by the new art of collage and assemblage, the emergence of which coincided with the rejection of craftsmanship on the part of the artist. In many cases, the arrangement of the studio and personal collection could now be seen as part of the creative outpouring of the artist, as a work of art in its own right.

Breton's atelier and personal museum

One of the most interesting examples of an atelier-home as an expression of its owner's creative imagination was the apartment at 42 rue Fontaine, at the foot of Montmartre in the Pigalle district of Paris which, for 44 years, was home to Breton.[13] Here Breton lived, wrote his works, made his art, and assembled a large and varied collection of objects which he always related to his philosophy of life. Despite the

The situational space of André Breton's atelier

evolution of the collection over the years, this "personal museum" remained the constant of the poet's life during three marriages and many other personal and artistic relationships. The apartment served also for many years as the "central office" of Surrealism, where social gatherings were held, poetic states entered into, exploratory games and surveys conducted, and Surrealist publications produced. The collection and its arrangement developed in parallel with the Surrealist philosophy of the found object, which in turn was rooted in Breton's passion for collecting. When Breton's apartment was stripped and the collection tragically dispersed in 2003, many felt it to be a ruthless act of cultural vandalism.

Breton's atelier was situated in two adjacent nineteenth-century studio buildings, entered through a courtyard off rue Fontaine, a street which, as Mark Polizzotti put it, "provided the short link between the nightclubs and brothels of Places Blanche and Pigalle, and the residential working-class modesty of rue Notre-Dame-de-Lorrette."[14]

Figure 17.3
42 rue Fontaine, Paris. a. Entrance from rue Fontaine; b. Courtyard side (Breton's windows on third level); c. Staircase B; d. North facade on Boulevard de Clichy (Breton's windows on third level in the two buildings on the left)

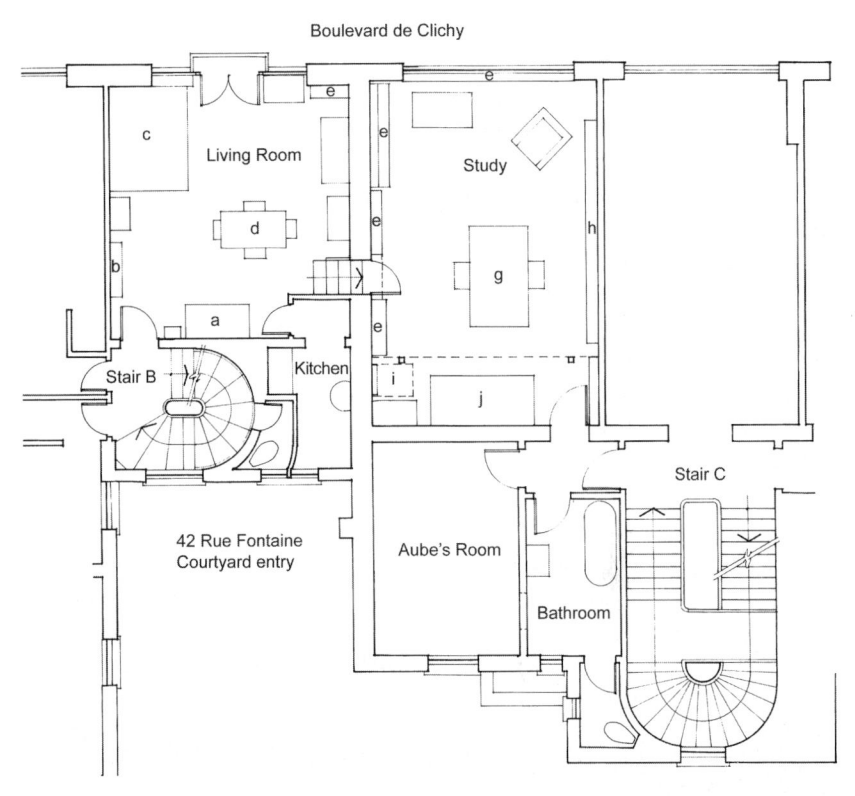

The situational space of André Breton's atelier

Figure 17.4
André Breton's apartment at 42 rue Fontaine: Floor plan: a. wardrobe; b. case with birds; c. double bed; d. dining table; e. bookcase; f. small table; g. Breton's writing table; h. the "wall" assemblage; i. fold-down ladder to mezzanine above; j. day bed

A Bohemian district since the nineteenth century, the area had been a gathering place for poets and artists, a fact which, together with its affordability, must have appealed to Breton and his friends. The 9th *arrondissement*, together with the Montmartre district immediately to the north, is one of the main areas in Paris for high concentrations of artists' studios. Access to Breton's apartment from the street was through a long, dark corridor of a low building housing a theatre (the Comedie de Paris). Walking past the stage door, one came into a deep, irregular courtyard, containing the *concierge's* lodge and overlooked by flats and artists' studios. The apartment was situated at the far end, facing the Boulevard de Clichy. It was on the third floor and accessible by one of two staircases belonging to different buildings. The one which the Bretons normally used, staircase B, was a narrow and somewhat ramshackle wooden spiral stair which when climbed, according to the poet, "gave the impression that anything was possible".[15] The landings were rather tight, each with three matching green doors. The modest apartment consisted of two main rooms, both filled with the objects and paintings of the collection: the living room and Breton's slightly larger study. Both these rooms had been built as double-height artists' studios, with expansive atelier windows facing north to the broad, tree-lined Boulevard de Clichy. These also overlooked Place Blanche and several cafés and nightclubs, including the famous Moulin Rouge. The living room served a variety of functions. Behind a heavily armoured entry door, it contained a small dining table, a double bed and a low enclosure leading to the kitchen, all surrounded by the collection. Access from the living room to Breton's adjoining study was up five steps and through a thick party wall, accentuating the room's status as a raised place of contemplation. Breton's study seemed like a cross between the dusty office of a naturalist and the disordered storeroom of an ethnographic museum. From his desk, Breton could see through and down into the adjacent salon. A visitor looking from the salon up to the study, would perhaps glimpse the poet at his enormous writing table, partly obscured by the piled up books, papers and objects, against the splendid backdrop of his "wall", the most elaborate, wall-size assemblage of paintings and objects in the atelier. On the inner side of the study was a shallow gallery typical of artists' studios. It was supported by slim columns and accessed by a fold-down wooden ladder in the corner. Below this gallery, near Breton's writing table, hung for a long time one of his most prized paintings, de Chirico's *A Child's Brain*. The courtyard side of the apartment contained a very small kitchen, a hallway connecting to the other staircase, and a bathroom decorated, with characteristic irony, with dozens of miniature ceramic folk *bénitoires* or holy water stoops.

Breton's collection was by its nature very private and often intimately autobiographical. Regarded as having personal significance, the objects were not labelled or catalogued. They were arranged and often rearranged with the ebb and flow of the author's imagination, and he often engaged in active dialogue with his pieces while working. In the principles of objective chance and "*dépaysement*", the object often acted as a stimulus to reflection and the imagination. The Surrealists' pursuit of such pieces was parallel and often equal in intensity to the pursuit of *l'amour fou* itself. Finally, the collection comprised equally both "works of art" (having a conventional market value) and common things, such as *trouvailles*, stones, used tickets and postcards – which only had personal value as stimulants and mementoes. Through their thematic arrangement, Breton's world was articulated.

Breton's personal museum, with its many natural specimens, owed much to the strange beauty of the collections of the natural history museums. Its thematic content and organizational structure, however, had much more in common with pre-modern collections, such as the traditional cabinets of curiosities. The *Kunst-* or *Wunderkammern*, and *studioli*, which thrived in the European culture of curiosities of the sixteenth and seventeenth centuries, were private collections of *naturalia* and *artificialia*, containing wonders of nature, scientific instruments, treasures of art and exotic, bizarre or occult rarities.[16] These objects possessed a magical, fetishistic quality for the collectors, having an affinity perhaps with that of the holy relics which preceded them in this tradition. They were grouped together *thematically* within the space of a specially designed room or container in a way which revealed their latent resemblances and contrasts. Among the goals of such cabinets was to represent the cosmos in miniature, and to show the power and erudition of the often princely collector. Their primary function was symbolic and cosmological: to inspire wonder in the beholder by revealing, through a web of analogies, the mysterious correspondences in the world between man and nature, the microcosm and the macrocosm. The same sense of mystery and wonder would permeate the Surrealist collections. One of the most interesting, already transitional cabinets in the Parisian context was the famous early eighteenth-century collection of Joseph Bonnier de la Mosson. Dismantled and partly reconstructed (as a kind of three-dimensional collage) in the royal gardens, it was on public view in the Jardin des Plantes until 1935, where the Surrealists would no doubt have seen it.

The situational space of André Breton's atelier and collection

In the light of the preceding discussion, we may now consider the spatial structure of Breton's personal museum. With an interest in the primitive ordering of things found in psychotic art, and in the *collagiste* sensibility of tribal art, Breton grouped things together organically, through the process of analogy and provocative juxtaposition. The most elaborate part of the collection was the large "wall" behind Breton's writing table in the study, on which I would like to concentrate here. Breton was sensitive to the careful relative placement of things, and the "vitally important *concrete relationships* that can [thus] be established between immediate objects".[17] In other words, he was sensitive to the power of the spatiality of his collection as a situational structure. The relationship of the wall to both the public rooms was meaningful.

The objects of the wall were grouped in loose, asymmetrical configurations, around a geometrically regular, stepped wooden shelving unit. It resembled a secular version of a Byzantine *iconostasis*: an ideal, grid-like structure, used symbolically for the veiling and revelation of cult mysteries, around which many images or fragments could be ordered. Breton's structure was approximately 4.5 m long × 1 m high × 25 cm deep, dividing the space of the wall into bays or compartments. The symmetrical framework, with its tripartite structure, was reminiscent of the centralization and geometric regularity of the sacred architecture of various religious traditions. Its tripartite form was reinforced by the placement of some of the objects, with three large paintings (Francis Picabia's *LHOOQ*, 1919; Joan Miró's *Head*, 1927; and Jean Degottex's *Black Pollen*, 1955) hanging above the three zones, separated by two similar, wood-carved ceremonial boards from New Guinea. The central bays of the shelving were enclosed, forming two niches. The structure thus established several places of particular importance. In the middle of one of these, in the longer central niche, Breton placed a framed photograph of his wife of the post-war years, Elisa. Her central position in the wall reflects her significance for Breton's life and work. He and the Chilean Elisa Claro had met in New York in 1943 in the aftermath of marital break up and much personal pain on both sides. Their marriage is sometimes portrayed as something of a salvation for each of them.

The situational space of André Breton's atelier

Figure 17.5
a. Breton's "wall" behind the desk in the study; b. Diagram of wall assemblage

The three zones below the shelving were each occupied by a different cabinet. On the right, near Breton's table, stood a configuration with vaguely anthropomorphic overtones.[18] At its base was an ornate, lacquered and gilded antique Spanish cabinet, above which stood the glass dome holding colourful Mexican birds, frozen in time. This vitrine corresponded to a large rectangular glass case with exotic birds near the entrance of the apartment and visible from Breton's table. The vitrine was flanked by two New Guinea over-modelled sculls, and surmounted by a fish-tooth and bead ceremonial headdress from the Marquesas Islands, one of many pieces of tribal body ornament in the collection. Above the shelf hung a crescent-shaped wooden pectoral from Easter Island, surmounted by a whale ivory and braided-hair pendant from Hawaii. The whole could be seen as an Arcimboldesque figure with a smiling face. Indeed, this cabinet and bird vitrine had been exhibited by Breton in the 1938 *Galerie des Beaux Arts* Surrealist exhibition in Paris equipped with a female mannequin arm and legs, under the title "the exquisite corpse".[19] The startling assemblage combined quasi-religious imagery with overt eroticism, highlighting the ambiguities of Catholic iconography. It also alludes to the Surrealist love of the corporeal, and to the power of physical fragments to conjure up wholes.

The central section of the wall consisted of a dark wood coin cabinet, surmounted by Alberto Giacometti's *Suspended Ball*. Behind it in the upper niche, at seated eye level, was the framed photograph of Elisa. If Giacometti's subtly erotic sculpture can on one level be read as pertaining to the interplay of male and female principles, of the sun and the moon, its proximity to Breton's wife suggests the thematics of the alchemical reconciliation of opposites permeating much of Breton's work. The photograph was surmounted by Henri Rousseau's 1907 *Still Life with Cherries*. This small picture of fruit and flowers was placed at standing eye level, among masks and over-modelled sculls, and below Joan Miró's large spectral *Head*. In this position, the still life recalled an offering in the shrines of Roman houses, thus hinting at some arcane ritual of a private cult.

The central positioning of Elisa on the wall merits further reflection. In the Romantic role of the muse, guide and "saviour" of the poet's later years, Breton's wife assumed here the position of the deity presiding over his life and home, complicit in his creativity. Her prominent placement in the wall is consistent with the Romantic theme of the paradigmatic woman, contemplated by the creative male.[20] It is further illuminated by the doctrine of the ideal feminine which Breton developed in the text of the early years of their relationship, while the world was still in the shadow of war, *Arcanum 17*. While the theme of the woman muse was a constant throughout his life's work, in this text his view of feminine power reached a new high. Through the ancient legend of Melusina (a magical half-woman, half-ondine creature), Breton argued that woman transcends the moral bankruptcy, exposed by the war, of the masculine world. In her more intimate, intuitive connections to regenerative nature, she is wise and semi-divine, and alone able to bring about the needed cultural resurrection.[21] This view of the feminine is related to the Romantic artist's view of woman as a kind of Dantean Beatrice figure, inspiring and guiding the poet's soul to paradise. Adorned by emblems of primeval power, she occupies the place of honour in Breton's wall as a kind of benevolent goddess of the temple. To the right of the framed photograph stood a glass box with a colourful bird (a medieval emblem of the soul), and then Breton's *Souvenir of Earthly Paradise*, a mounted and inscribed stone which he found in southern France in 1953.[22] The stone's title, particularly poignant since the poet's death, is expressive of the timeless metamorphic processes of nature, and of the sensual joys of being alive. It seems appropriate that this object should have been positioned in close proximity to Breton's companion of later years, and one of the chief custodians of his personal museum after his death. This grouping is a good illustration of the seamless blending of the cosmic and the deeply personal which characterized the work of Breton and many modern artists. The concrete configuration of the objects, grouped together with a sensitivity for their mutual relationships, structured a rich communicative space.

Conclusions

In my discussion I have focused on the thematic field inherent in the situational spatiality of a portion of Breton's study. I have tried to show that the individual identities of the pieces are insignificant in comparison to their role within the situational structure of the room: their mutual spatial and thus also thematic relationships within the whole. This emphasis on the multifarious whole in which each element is embedded is at odds with the attitude of many art historians and gallery curators, whose instinct might be to select the "best" and most valuable pieces for display. In his collection, Breton consciously sought to redress the sterile isolation of the art object in conventional museums, providing a situational context for each thing, whether conventionally valuable or not. The whole collection, with its personal and idiosyncratic elements, has to be seen as a "mapping of the landscape of its author's imagination".[23]

A similar preoccupation with the thematic potential of the fragment is evident within the work of some of the twentieth-century's best architects. Le Corbusier, for example, recognized the technique of juxtaposition and situational grouping as a quintessential expression of a modern sensibility,[24] deploying it in the many forms of his creative endeavour to create rich thematic worlds. Carlo Scarpa raised the fragmentary in his architecture to a meditation on the spectral presence of the past within daily life, while also thematising the subtle and surprising junctions and relationships of elements within his architecture.

Finally, a consideration of Breton's atelier brings into focus the museum-like nature of contemporary culture. Although his own collection must be understood as a critique of the decontextualization of works within the conventional museum, it nevertheless manifests the problematic need in the twentieth century for an introverted, private culture, gathered within the private home-museum. In this vision, the traditional public and ethical realm of the city is reinterpreted as an agglomeration of private domestic worlds.

Notes

1. See Dalibor Vesely in "Architecture and the Ambiguity of the Fragment", in R. Middleton (ed.) *The Idea of the City*, London: The Architectural Association, 1996.
2. This chapter is drawn from the longer essay: Dagmar Motycka Weston, "Communicating Vessels: André Breton and his Atelier, Home and Personal Museum in Paris", *Architectural Theory Review*, Vol. II, No. 2, 2006, pp. 101–28.
3. See Debora Silverman, "A Fin-de-Siècle Interior and the Psyche: The Soul Box of Dr Jean-Martin Charcot", *Daidalos* 28, 1988, pp. 24–31.
4. Edmund Engelman, *Berggasse 19: Sigmund Freud's Home and Offices, Vienna, 1938: The Photographs of Edmund Engelman*, New York: Basic Books, 1976.
5. See Hilda Doolittle, *Tribute to Freud by H.D.*, Oxford: Carcanet Press, 1971.
6. John Golding, "Picasso and Surrealism", in R. Penrose, J. Golding (eds) *Pablo Picasso 1881–1973*, Ware: Wordsworth, 1988, p. 96; Breton, *L'Art magique* (c. 1957), Paris: Adam Biro, 1991.
7. Figures such as Marcel Mauss and Lucien Lévy-Bruhl, and later Claude Lévi-Strauss, Michel Leiris and Roger Caillois, examined the thought patterns among non-Western and "primitive" peoples, especially with respect to the magical and symbolic significance of all aspects of their reality.
8. Claude Lévi-Strauss, *The Savage Mind*, London: Weidenfeld and Nicolson, 1966, p. 9.
9. Hans Georg Gadamer, *The Relevance of the Beautiful and Other Essays*, Cambridge: Cambridge University Press, 1986, p. 7; Peter Carl, "Le Corbusier's Penthouse in Paris: 24 Rue Nungesser-et-Coli", *Daedelos* 28, 1988, pp. 65–71; and "The Godless Temple, organon of the infinite", *Journal of Architecture*, Vol. 10, No. 1, 2005, pp. 63–90.
10. See for example Michael FitzGerald, *Picasso. The Artist's Studio*, exhibition catalogue, New Haven, CT: Yale University Press/Wadsworth Atheneum Museum of Art, 2001.
11. See A. Breton, *Surrealism and Painting*, Boston, IL: MFA Publishers and New York: DAP, 2002, pp. 363–66.
12. Breton, "Picasso dans son élément", *Minotaure* 1, June 1933, pp. 4–28; M. Reynal, "Dieu-Table-Cuvette", *Minotaure* 3–4, December 1933, pp. 39–53.
13. I am grateful to the present *concierge* and tenants of 42 rue Fontaine and to Agnès de la Beaumelle for their help.
14. Mark Polizzotti, *Revolution of the Mind: The Life of André Breton*, London: Bloomsbury, 1995, p. 167.
15. Breton, letter to Marianne Oswald (1959), in *André Breton 42 rue Fontaine*, p.5.
16. See Dagmar Motycka Weston, "'Worlds in Miniature': Some Reflections on Scale and the Microcosmic Meaning of Cabinets of Curiosities", *ARQ*, Vol. 13, Issue 1, August 2009, 37–48.
17. Breton, *Surrealism and Painting*, p. 12 (original emphasis). He was speaking in this text with reference to Cubist collage.
18. See Gracq, *42 rue Fontaine*, Pl. VIII and XIV.
19. Lewis Kachur, *Displaying the Marvelous*, Cambridge, MA: The MIT Press, 2001, pp. 71, 75 and 77.
20. This theme is noted by Peter Carl in "The Godless Temple", p. 63. When Breton faced the wall to write, the picture of Elisa would have been at his eye level.
21. A. Balakian, "Introduction", *Arcanum 17*, pp. 12–17.
22. Breton, *Je vois j'imagine: poèmes-objets*, Paris: Gallimard, c. 1991, pp. 126–29.
23. This expression is a paraphrase of the great Czech surrealist film-maker Jan Svankmajer's description of his own surrealist *Wunderkammer* in southern Bohemia. Conversation with the author, May 2007.
24. Le Corbusier, Pierre Jeanneret, *Oeuvre complète 1934–1938*, Zurich: Girsberger, 1939, p. 157.

Chapter 18

L'histoire assassinée

Manfredo Tafuri and the present

Teresa Stoppani

A new attention

In recent years the work of Italian architectural historian Manfredo Tafuri has attracted a lot of interest in the world of architectural theory and practice. After the publications that celebrated Tafuri's work immediately after his demise (most notably *Casabella*'s "The Historical Project of Manfredo Tafuri" in 1995), or produced a first collective critical reconsideration of his legacy shortly thereafter (with the seminal issue of *ANY* entitled "Being Manfredo Tafuri" in 2000), the last few years have seen the publication of the English translation of Tafuri's last book on the Renaissance and of new studies on his work, which in different ways have returned Tafuri's legacy to the forefront of the architectural debate.[1]

The interest in Tafuri's work though had never entirely died off. Beyond the immediate applications of his historiographical method by colleagues and students at the Department of Architectural History at the Institute of Architecture of the University of Venice (IUAV), who continued his investigations in certain areas of research, producing, for instance, a vast series of studies on the architecture of the Renaissance, and beyond the stimulus offered by his critical historical "project" to recent North-American architectural theory, Tafuri's history as an open "project" addressing the crises of architecture has remained a constant point of reference also for the architectural project.[2]

Critical history

The recent books on Tafuri reconsider his work not only in terms of its relation with the history of a remote architectural past (the Renaissance of his late studies), but also in the ever difficult relationship of history with the present of the discipline. The history of the present, and particularly the relationship of history with the present of the architectural project, is the main focus of Tafuri's early writings. In texts that range from *Theories and History of Architecture*, to *Architecture and Utopia* to *The Sphere and the Labyrinth*, Tafuri redefines the role and the method of architectural history as a never-ending "project", an open and self-questioning process rather than a finite and defining story. Tafuri wants the historical "project" to be separate and autonomous from the architectural project, in order to provide the tools to return, independently, to the architectural project, and address it with a critical analysis. It is in this sense that Tafuri provocatively declares that criticism does not exist and only history does. History is necessarily critical and can only be critical; on the other hand, a criticism without history, without the analytical tools and the distance of history, does not exist, as it cannot be critical without it.[3]

History of the present

Tafuri's key legacy is the definition of a history of architecture that is both differentiated from the history of art and detached from the project of architecture. Situated in the difficult space of this distance, presented as an argued provocation, the history of architecture thus redefined is precariously constructed as an autonomous and yet always compromised discourse that proceeds in parallel with architecture. Like architecture, it is also open and exposed to the forces of "multiple techniques of environmental formation".[4] The crucial test of this history (and indeed of all histories) takes place in its relation to the present, and yet, even when Tafuri's research addresses the remote past, its objects are never quiet and appeased, but always complex scenarios of crisis, breaking points, sites of ambiguities, moments of shifts in mentalities and power relations. The history of the past remains open, bearing questions and unresolved problems that find echoes in the present, and that the present needs to address; the present becomes a complex and complicated system of relationships rather than a singular and circumscribed moment.

Tafuri's own present is the time of the reconsideration of the effects of the crisis of modernism on contemporary architecture; it is also the time of the crisis of the architectural language, within the definition of a long modernity that finds it roots in the Renaissance and in its manipulations of classical languages. It is, finally but not conclusively, the present in-progress of the evolving definition of the historical "project". For these reasons, for its openness to both the past and the future, the historical "project" remains a present issue and a work to be continued.

L'histoire assassinée: Manfredo Tafuri and the present

History as research

Tafuri's last book *Ricerca del Rinascimento*, published in 1992 in Italian, was recently published in English under the reductively rendered title of *Interpreting the Renaissance*.[5] The book is a specialist study of the complex relationships between the urban architecture of the Renaissance and its physical, cultural and political contexts, and focuses in particular on the power systems that affected (then, as they do now) architectural production. Important for its innovative interpretations of specific buildings, projects and urban environments, Tafuri's work constructs a history to which the present cannot be indifferent. In his methodological and critical preface Tafuri relates the formal research of the Renaissance to the more recent uncertainties and moments of experimentation of the modern avant-gardes, explicitly linking an "unresolved past" to the "unsettled present."[6]

The "research" of Tafuri's original title is twofold and bidirectional, tense and dynamic. "Research" is the architectural research of the Renaissance, a difficult, problematic, discordantly polyphonic, heterogeneous research that took on different religious, political and architectural instances, and progressed at different and uneven speeds. "Research" is the historical research ("project") conducted by the historian on material that is still partly hidden, partly invisible, often ambiguous and susceptible to different interpretations. This is a research that does not restrict itself to the presentation of the past, but links it to the present, addressing issues that remain topical today in the complexity of the political decision-making processes, as well as in the compromises, social implications, criticism and choices of languages that architecture always deals with.

The past of Tafuri's Renaissance constantly challenges the present with its unresolved problems, unsettling the very role of the historical research on it. There is therefore a third level of research that is the true object of this book. Beyond the research conducted by the artists, architects, intellectuals, princes and politicians of the Renaissance, and beyond the research produced today on these by the historian, there exists a third level of an exquisitely Tafurian research on the tension that relates the two. This is a research on the discipline of architecture, and on the reasons why it is necessary today (or nearly 20 years ago, when Tafuri's work was first published), to consider the problem of the crisis of architectural languages incubated in the Renaissance, in order to understand the questions of contemporary architecture after the crisis of modernism.[7]

History and the present

Tafuri's relationship with the present, both in the definition of a historical methodology and in his critical writings on his contemporaries, was never simple. While Tafuri's history of the present strives to remain independent from architectural practice and produce a critical distance from design, Marco Biraghi has also identified the possible influences that Tafuri would have shared with or derived from architectural practice.[8] Diane Ghirardo has examined the influence of Tafuri's work on architectural theory and design research in the United States from the 1970s to the 1990s, concentrating on the reception, appropriation and misunderstanding of Tafuri's thought by a few contemporary American architects, in particular Peter Eisenman and Daniel Libeskind.[9] For Ghirardo, the self-referential architecture developed by Eisenman and Libeskind in the 1970s and 1980s represents a retreat from the architecture of political and ideological engagement that Tafuri had advocated in his historical and theoretical manifestos: "three decades of theoretical delirium in which poeticizing reflection passed for theory".[10] When she refers to architectural theory, Ghirardo does not distinguish between texts and projects, considering a theory of architecture that is developed and proposed through both texts and design proposals. In a way Ghirardo continues to acknowledge the existence, parallel to the historical discourse, of an "operative criticism" performed by architecture onto itself.[11] The problem of the positions in architectural design and theory that Ghirardo analyses is that they do not relate to Tafuri's work from a distance, through distinctions and oppositions, as the architectural (design) research of the 1960s had done. Eisenman and Libeskind take on Tafuri's positions and incorporate them in the service of the architectural project – and here lies the problem.

Architecture and the historical "project"

At this point it is necessary to define what I mean here by architecture, before moving to a consideration of Tafuri's relation to the architecture of the present, his varied "present" from the 1960s to the early 1990s of an Italy in fast transformation. I call here "architecture" – to distinguish it from architectural history – that thinking, making and producing of architecture as a critical and self-reflective practice that may or may not be directly involved in, or productive of, the construction of physical environmental artefacts. In this sense, architecture at large is to be intended beyond the practical contingency of professional practice as a system of production. The two partially overlap, but they do not coincide. "Architecture", as inclusive of critical considerations of its role in the making of space and in the definition of its own languages, may often produce effects and shifts (as it did in the period that I consider here, from the 1960s to the 1980s) that do not find immediate and direct application in the built environment, but continue to occupy the space of the discipline, and only slowly filter through it, in time, changing and adapting it. "Architecture" thus defined is the space for the production of ideas in relation to design, for the experimentation of architectural languages, for the consideration of the role of architecture in the city and in society at large, and for a constant self questioning.[12]

In relation to this space, Tafuri's historical "project" occupies a fundamental role, uncomfortably offering to architecture an "other" critical conscience that, far from offering solutions, operates through a never appeased investigative attitude. Tafuri's analysis continues to explore and challenge itself as well as its objects, constantly redefining, in the process, its own role, questions and methods. It is an unsettling and challenging voice, which remains always inquisitive and self-critical in its provisional and precarious constructs. Historical investigation is for Tafuri not a form of criticism, but a critique of ideology: it does not linearly narrate and support a certain orientation, language or movement in architecture (what "operative criticism" does), but constructs a problematic, independent and open critical history that challenges both the past and the present.[13] In *Theories and History of Architecture* Tafuri had analysed the established methodologies of the history of architecture and had defined "operative criticism" as that form of manipulated and intentionally biased history constructed by critics who were also active in architectural practice, or actively engaged in support of specific trends in design, an instrumental history. "*Operative criticism* ... has as its objective the planning of a precise poetical tendency, anticipated in its structures and derived from historical analyses programmatically distorted and finalised."[14] Tafuri traces the beginning of operative criticism in Giovanni Pietro Bellori's seventeenth-century *Vite de' pittori, scultori et architetti moderni* and then identifies it, among others, in Sigfried Giedion (*Space, Time and Architecture*) and in his Italian contemporaries Bruno Zevi (*Storia dell'architettura moderna*) and Leonardo Benevolo (*Storia dell'architettura moderna*). Opposed to these biased constructions of history, Tafuri's works, especially the texts which introduce a method for the definition of an independent architectural history, remain an always uncomfortable and challenging voice for architectural practice.

A live(ly) debate: Tafuri and Rossi

A multiple question opens up here. What is it of the historiographical and historical work of Tafuri (that is, of his methodological definition and application of the historical "project") that *imports* to architecture and to the architectural project? This question implies a mutual reactivity between Tafuri's critical history on one hand and a critical architectural practice on the other. In other words, it asks: What of Tafuri's work is important and relevant to architecture? What of it addresses and affects the architectural project from the outside? It also asks: Beyond influences and tense conversations, what of the history as a "project" that Tafuri proposes is appropriated by architectural practice, often at the risk (as Ghirardo has shown) of distorting and voiding its message?

The answers to these questions would require a series of articulated investigations. Here I consider only a specific moment of Tafuri's complex relationship with contemporary architecture by focusing on his position, as a critic and historian of the present, on some aspects of the work of Aldo Rossi. Tafuri and Rossi operated in the same years, city and institution, developing their research and their teaching within the IUAV; yet they lived in the very "distant" and different worlds of the Department of History of Architecture and the Department of Architectural Design. The departments and their protagonists in the period of time that runs from the 1960s and throughout the 1980s are internationally known as the "School of Venice", a general spatio-temporal label that embraces all the different groups, positions and movements active at the IUAV at the time, encompassing all the tensions, conflicts and mutual references, attacks and collaborations that animated the dialogue in the most intense and critically productive moment in the history of the institute.[15] In that context, the relationship between the work of the historical "project" on the present, and the work of the architectural "project" in the very same present was more active than ever. The object of history here was the present, and the architectural present replied and reacted to its contemporary critical history in the first person. Architecture responded with its own tools – the image, the drawing, but also the text – to the provocations of critical history; history counter-attacked with its words. Here the historical "project" as interminable analysis engaged a difficult object that was not only changing and responsive, but was also in itself structured to be a self-critical process on the architectural language in its complex relation to the urban scenario past and present.

L'architecture assassinée: Research as figure

Aldo Rossi's watercolour *L'architecture assassinée*, dedicated to Manfredo Tafuri, is emblematic of the relationship between the two. A visual response to Tafuri's denunciation of the crisis of architecture, its languages and its social and political engagement expressed in *Architecture and Utopia*, Rossi's image shows his own architectures of pure geometric solids, urban typologies and personal memories broken into pieces, fractured and collapsed. The breakings that Rossi represents are in fact at the core of the relationship between Tafuri's historical "project" and Rossi's critical architectural "project". Both produce breakings, in the language of architecture and in the methods of history, in order to communicate, in different ways, a grounded criticism of architecture, from within the project (Rossi) and history (Tafuri). The breakings that Rossi draws are also a symptomatic representation of a shift in his own work towards abstraction (from the reality of the city), the analogical (of a city reduced to figure) and the formal (of a self-referential and obsessive personal language). They also mark the breaking that such shift produces in Tafuri's relationship to Rossi's research, and his critique of it. The time lag and the space between languages occupied by the translation of Tafuri's book *Progetto e utopia* into its English version made it possible to present Rossi's reaction to Tafuri's text, together with the text itself, in a clever editorial move of critical synthesis. *L'architecture assassinée*, Rossi's rebuke in drawing to Tafuri's suggestion that architecture as a project is "dead", becomes the cover image of the MIT Press edition of *Architecture and Utopia*.

The conflicting and changing relationship between Tafuri and Rossi, characterized by clear contraposition as well as unresolved ambiguity – can be well

L'histoire assassinée: Manfredo Tafuri and the present

Figure 18.1
Manfredo Tafuri's *Architecture and Utopia*

illustrated by an analysis of two texts by Tafuri on Rossi's works. Published nearly 20 years apart from each other, the texts well summarize and clarify Tafuri's position on the evolution of Rossi's architectural "project". In *Teorie e storia dell'architettura*, Tafuri considers Rossi's "silent architectural objects" as the effective evidence of the merging of architectural criticism with the criticism of the city. For Tafuri this combination results from "the wish to adhere with enthusiasm to the multiple pressures of urban reality and, at the same time, to introduce in it *architectural events* and fragments which might force the entire meaning of that reality".[16] The project of architecture and its experimentations are not limited to questioning and recomposing the language of the architectural object, whether this is derived from the language of a rationalist modernism, or from a typological history of architecture, or from personal memories and suggestions; all of which are combined in Rossi's case. The project enacted by Rossi's "*architectural events* and fragments" is in fact urban. Partial and fragmented, Rossi's projects renounce the control of the urban plan; locally inserted in the city, they perform a critical architectural act that goes beyond the city, offering in fact a critique of criticism through a drawn and built architecture. Tafuri reads Rossi's projects in the city as a form of *architectural* criticism, that is, criticism performed *in* and *by* architecture.[17] This is not a project of history, and it is from this sort of operation that Tafuri wants to establish a clear distance. But while the criticism of the "critics" can only be dismissed, Tafuri remains interested in his alter-ego Rossi, who constructs a critical project in architecture initially similar, or parallel, to Tafuri's critical project in history. Fragmented and partial, Rossi's work operates rigorously on its own language, suggesting a method to develop a language able to address the past of the discipline, as well as the present of the city (and the past in it). These are the aspects of Rossi's work that interest Tafuri: Rossi's project goes well beyond its suggestive images and evocative memories, offering a critical architectural tool that is always productive of instability. The collapses portrayed in *L'architecture assassinée* then are not caused by Tafuri's attacks on Rossi, but are in fact already genetically imprinted in Rossi's architecture, designed to operate like a surgical instrument on a city that it does not control. It is this critical (and self-critical) element of Rossi's project that Tafuri respects from the distance of history. Not only critical of its own language and critically acting on the city, this architecture is also critical of criticism. "[A]rchitectural criticism" writes Tafuri, "puts in crisis the critics of architecture. On the contrary, since the traditional task of criticism is already realised within the architectural structures, one could say that an independently critical architecture has the objective of destroying any critical intervention from outside."[18]

L'architettura della città: Research as architectural text

Tafuri describes Rossi's studies on the city, mainly *The Architecture of the City* but also the series of essays produced during Rossi's academic career, as a form of "operative criticism" that is both typological and urban.[19] For Tafuri, Rossi's work is a form of criticism because it is grounded on extensive research and on history. It is typological because it focuses on phenomena that remain "formally invariant".[20] Yet Tafuri's interest lies mainly in the fact that Rossi's typological criticism remains essentially urban and proposes planning decisions for the contemporary city, "even if only on the level of the structure of the image".[21] This form of criticism starts from a direct reading of the urban system, and is "immediately translated into systems that [modify] its single components, or, in extreme cases, its fundamental laws".[22] Rossi suspends his judgement on the city as a whole, in favour of "the analysis on limited sectors-environments, that are seen [as] the most vital aspects of urban structure".[23] For Tafuri the important feature of this approach is its unsettling work on the established city, on which it produces formal experimentations as critical analyses that "upset, examine, reassemble in new forms, the structural elements that the contemporary city tends to see as immutable and undisputable values".[24] What is essential, in this experience, is that "historical analysis, critical examination, critical function of the image and demonstrative value of planning, are all indissolubly connected" (160). Here resides the critical strength of Rossi's research: beyond the isolated formal exercises and the manipulation of elements abstracted from the city – real, remembered or imagined – Rossi's work addresses the dimension of planning. Manipulating, breaking and endlessly experimenting, Rossi refuses the "a priori existence of well-defined form" and a "precise configuration for structures", ultimately questioning the formal dimension of planning.[25] Rossi's projects and writings are not history; they are, at most, a form of "operative" criticism, and it is their critical dimension that interests Tafuri. Produced in architecture and through its forms, Rossi's "projects" (both his architectural designs and his writings on architecture) approach history as an inheritance to be challenged, questioned and redefined. For Tafuri, what is of import is the attention that Rossi pays to the transformations of the physical and anthropo-geographical environments, and his attempt to understand "the meanings underlying the transformations" through "the partial questions asked by architecture of architecture".[26]

Analogy: Figures without research

Later works by Aldo Rossi seem to lose the critical energy of his early typological and urban investigations, retracting into a personal sphere of speculation where forms are iterated in a sort of private mantra of memory, oblivious of any relation with the real city. The plate *The Analogical City* and the proposal for the *Roma Interrotta* workshop in 1978 are examples of such involution in Rossi's work.[27] The city here is reduced to a figure loaded with architectural evocations and personal memories, but removed from the real city, its physical structure, its society and its planning. These projects for the city remain repetitions of known forms and architectural references, the aggregation of which does not raise interrogatives on the making of the urban space, and even less on the meanings of such making. It seems obvious then that Tafuri should be strongly critical of Rossi's later work, where the image and its combinatory proliferations are only figures of a subjective "poetics", divested of the analytical and critical attention to the real city that Tafuri had identified in Rossi's early research. Referring to Carl Jung's definition of "analogical" thought as "archaic, unexpressed, and practically inexpressible in words", "an interior monologue" opposed to "logical" thought that is "expressed in words [and] directed to the outside world in the form of discourse", Rossi's work retracts in an aphasic project that invests the image with analogical meaning.[28] Rossi presents this move as a step forward in an interrogation of architecture by architecture, capable of reaching where the words of "logical" discourse cannot reach. In the same years of his critique of Rossi's painted and collaged analogical architecture, Tafuri's examination of Rossi's built projects recognizes in them a corresponding waning of criticality in relation to architecture and to the planning of the city.[29] The publication of *Storia dell'architettura italiana 1944–1985* gives Tafuri the occasion to reconsider Rossi's research in retrospective, nearly 20 years after the theoretical and methodological framework that he had identified in *Teorie e storia dell'architettura* in the 1960s and more than ten years after the architectural "collapse" depicted in Rossi's *L'architecture assassinée*.[30] In this book Tafuri reads the Gallaratese housing complex planned by Carlo Aymonino (1967–1970, built 1970–1973) as an implementation of the strategy of the formation of the city "by finite parts", but an over-designed "monument to noise" incapable of "'putting order' in the ocean-like periphery of the metropolis".[31] Part of the complex, Rossi's block acts as a "hieratic and sustained" silent witness of Aymonino's theatrical *mise en scène*. For Tafuri this building, like other Rossi works from the same years, shows that by now his research "resists any compromise with reality, because the return to the 'ancient house of language' is possible only through an affirmation of aloof indifference'."[32]

For Tafuri, Rossi's early "architectural project", both in his designs and in his writings, had been a search for those primary forms "that are exiled from the urban space, but intend to speak of their exile, to propose a theory of the city as the locus of collective memory."[33] This search for form and research through form is what Tafuri had clearly identified as a critical work internal to architecture. In this sense the estrangement, the silence, the abstraction, the suspension of life in Rossi's early projects can be read as a stubborn construction, through architecture, of an enforced distance that is necessary for the project to perform its critical act. The congested amassing of forms in Rossi's drawings and paintings and the distillation of silent forms in his early built projects produce the same critical distance that Tafuri constructs in his "historical project". Even the demolition of forms and of their relations that is performed in *L'architecture assassinée* echoes the active, critical and distance-producing destruction that Tafuri employs in his project of history.[34] Destruction, criticality and distance characterize, up to this point (mid-1970s) and in different ways, the project of Tafuri in history (historical "project") and of Rossi in architecture (architectural "project"). Then, after the "assassination", what dies is not the critical history of Tafuri, nor the architecture of Rossi, but that critical research into the meanings of architecture that had made the latter a form of criticism (or, in Rossi's words, a theory of the architectural project). After the "assassination" performed by Rossi, Tafuri provisionally concludes, in apparent contradiction with his previous words in *Theories and Histories*: "The representation is all: it is pointless to strive to find in it hidden meanings in regions that it can not access. The city proves to be (…) a simple pretext."[35] In Rossi's projects of the 1980s Tafuri finds, "finally exhausted", the tradition of the critically operative urban studies that had been introduced in the late 1960s by Rossi and Aymonino. But he had seen it coming: for Tafuri the static destruction represented in *L'architecture assassinée* was in fact already "a frozen ruination: the fragments hanging or thrown into the void, remain still. This loss is not painful: the wayfarer was prepared for this".[36]

Notes

1. Manfredo Tafuri (1992) *Ricerca del Rinascimento. Principi, città, architetti*, Turin: Einaudi; Tafuri (2006) *Interpreting the Renaissance. Princes, Cities, Architects* (trans. Sherer, D.) London and New Haven, CT: Yale University Press; Marco Biraghi (2005) *Progetto di crisi. Manfredo Tafuri e l'architettura contemporanea*, Milan: Christian Marinotti Edizioni; Andrew Leach (2007) *Choosing History*, Ghent: A&S books.
2. Paolo Morachiello (1985) "The Department of Architectural History, 'A Detailed Description'", in Semerani L. (ed) *The School of Venice* (AD Profile), London: Academy.
3. Richard Ingersoll (1995) "There is no such thing as criticism, there is only history" (interview with Manfredo Tafuri), *Casabella* 619–20: 97.
4. Manfredo Tafuri (1987) *The Sphere and the Labyrinth. Avant-Gardes and Architecture from Piranesi to the 1970s,* Cambridge, MA and London: The MIT Press, pp. 1–2.

5   I discuss the translation of the book and of its title in my review of Tafuri's *Interpreting the Renaissance*: Teresa Stoppani (2008) "Book reviews", *The Journal of Architecture*, 13(3): 345–54.
6   "Passato irrisolto, inquieto presente" is the title with which *Casabella* publishes the preface of *Ricerca del Rinascimento*, anticipating the publication of the book. "Unresolved past" and "unsettled present" are also Tafuri's closing words: "The 'weak power' of the analysis, in other words, is proposed as a moment in a process that keeps alive the unresolved problems in the past, thus unsettling our present." Manfredo Tafuri, (1991) "Passato irrisolto, inquieto presente", *Casabella* 585: 40. Author's translation.
7   "Contemporary" is a term that Tafuri would most likely refuse, as for him the "contemporary" is still part of a long and far from resolved modernity. Tafuri would rather speak of the "present". Quite significantly then, Tafuri and Francesco Dal Co's *Architettura contemporanea* (Milan: Electa, 1976) is published in English as *Modern Architecture* (New York: Harry N. Abrams, 1979).
8   In particular, in the case of the Tafuri–Rossi relationship, Biraghi identifies as shared influences the thoughts of Karl Kraus and Walter Benjamin. See Marco Biraghi, 'Il frammento e il silenzio', in *Progetto di crisi*, op. cit., pp. 171–204 (and in particular pp. 184–204 on Tafuri and Rossi). In the introduction to his book Biraghi anticipates that for Tafuri "The real knot to untangle consists [...] in establishing what relationship links [...] the critical act to the 'given': that is, how much can the critical act 'detach' itself from the 'given', when it is ultimately 'made' of its same matter?" ("Il vero nodo da sciogliere consiste [...] nello stabilire quale legame vincola tra loro [...] l'atto critico e il 'dato': insomma, di quanto l'atto critico riesce a 'staccarsi' dal 'dato', se in ultima analisi è 'fatto' della sua stessa materia?") Biraghi, p. 19. author's translation.
9   Diane Ghirardo (2002) "Manfredo Tafuri and Architecture Theory in the U.S., 1970–2000", *Perspecta. The Yale Architectural Journal*, 33: 38–47.
10 *Ibid.*, 45.
11 Tafuri (1968) *Teorie e storia dell'architettura*, Bari: Laterza.
12 As early as 1966 Aldo Rossi had clearly articulated the manifold nature of the architectural discipline in his text "Architettura per i musei", published in the collective volume, Guido Canella (ed), *Teoria della progettazione architettonica* (1968), Bari: Dedalo; now available in Rossi (1975) *Scritti scelti sull'architettura e la città, 1956–1972*, Milan: CittàStudi, pp. 323–39. Rossi identifies theory's most important "moment" in the "relationship between the theoretical view of architecture and the making of architecture" ("il rapporto che esiste tra visione teorica dell'architettura e il fare architettura"), Rossi (1975), p. 323, author's translation. Rossi sees the "thinking" (*pensare*) in architecture as one with the "design" (*progettare*) of architecture. Subjective and far from the modernist "method", architectural theory for Rossi remains internal to the discipline of design and produces an autonomous discourse that does not need to refer to external disciplines (such as economy, sociology, linguistics).
13 Tafuri 1987, 1–21; Biraghi 2005, 9–53.
14 Tafuri 1980a, *Theories and History of Architecture*, London: The MIT Press, p. 141.
15 On the "School of Venice" see Luciano Semerani (ed) *The School of Venice* (1985) Architectural Design, 55. See also "Urban Planning: Giuseppe Samonà and 'The School of Venice'", in Margaret Plant (2002) *Venice. Fragile City 1797–1997*, New Haven, CT and London: Yale University Press, pp. 342–44. For a narrow definition of the "Scuola di Venezia" as it is intended in Italian architectural culture, see Pier Vittorio Aureli (2008) *The Project of Autonomy: Politics and Architecture Within and Against Capitalism*, New York: Princeton Architectural Press, note 12, pp. 84–85.
16 Tafuri 1980a, 130.
17 *Ibid.*
18 *Ibid.*
19 Aldo Rossi (1966) *L'architettura della città*, Padua: Marsilio.
20 Tafuri 1980a, 158.
21 *Ibid.*
22 *Ibid.* p. 159.
23 *Ibid.*
24 *Ibid.*
25 Tafuri 1980a, 160.
26 *Ibid.* 173.

27 Michael Graves (ed) (1979) *Roma Interrotta* (AD Profiles, 20), London: Academy, pp. 88–90; Aldo Rossi, (1976a) "La Città Analoga: Tavola/The Analogical City: Plate", *Lotus*, 13: 5–8.
28 Rossi, (1976b) "An Analogical Architecture", *Architecture and Urbanism*, 5, p. 74.
29 Pierluigi Nicolin (2000) "Tafuri and the Analogous City", *ANY*, 25–26: 16–20.
30 Tafuri (1986) *Storia dell'architettura italiana 1944–1985*, Turin: Einaudi.
31 Tafuri 1986, 152. Author's translation.
32 *Ibid.*, 166.
33 *Ibid.*, 167–68.
34 Tafuri derives his idea of productive destruction from Walter Benjamin's definition of the role of destruction in his philosophy of history.
35 Tafuri 1986, 170.
36 Tafuri 1986, 171.

Chapter 19

Nature choreographed

Anne-Louise Sommer

Aesthetic theories of landscape and garden

In the eighteenth century a veritable craze swept simultaneously over Great Britain, the Continent, Scandinavia and Russia. Influenced by the Enlightenment, a new tradition of garden design blurred boundaries between landscape gardens and the surrounding scenery. This new concept of garden aesthetic, often referred to as "the English garden" among many other names such as "the picturesque garden", "the informal garden" or "the irregular garden" will be examined through a tentative reading of the Pavlovsk Park in Russia. This eighteenth-century landscape park near St Petersburg places into perspective some of the challenges of research into the landscaped garden. Although contemporary writings discuss Anglomania to refer to this seemingly endless stream of scenes and modes of regarding the landscape as a stage that influenced the garden layouts, it can be questioned to what extent the source of this new garden aesthetic was solely British. There were prolific circles in Germany and France as well as in England. The underlying aesthetic ideas for the garden design of the period drew equally on philosophy, literature, painting and sculpture.

Latent within any garden design lies a mental image that becomes materialised in the layout. Within this ideational structure are embedded ethical values, aesthetic ideals and even political visions. Primarily the garden or the designed landscape is determined and defined by its materiality and its presence at an entirely tangible level, no matter how more abstract components will be dealt with, in respect to a privileged theoretical approach. Taking as a starting point the dialectic of poetics and physical form, one can readily see the great contribution of the humanities to a theoretical investigation of the process by which the eighteenth-century garden came into being.

When a recently published reader entitled *Theory in Landscape Architecture* from the University of Pennsylvania Press subsumes the contributions in examples of "instrumental", "interpretive" and "critical" theory it points to the fact that theories of the designed landscape in general are deeply rooted in specific professional traditions, and only to a lesser degree aims to provide an interdisciplinary access.[1] Discussions of landscape architecture often offer questions of soil analysis, conditions of growth or quality of foliage as pivotal points of debate on one hand, or more subtle philosophical issues on the other. In between, a variety of different approaches regard the designed landscapes purely as a work of art in line with an art history tradition, or, from a different point of view, a way to focus on ideological aspects rooted in socio-economical insights. Each one of these approaches provides only a limited understanding of the wide-ranging potentials of the garden art and landscaped design as an object for interpretation.

In the foreword to the English edition of *Les Mots et les choses* Michel Foucault writes: "Discourse in general, and scientific discourse in particular, is so complex a reality that we not only can, but should, approach it at different levels and with different methods."[2] In line with Foucault's suggestion of the discursive multivalent strata available, garden art and designed landscapes may be considered to possess a critical dimension reflective of and responding dynamically to the epistemological systems of a given period. The unique potentials of nature as opposed to culture in the landscape garden might even argue for a privileged and ideological position, at once formative for the discursive field and in critical opposition to the very same, giving rise to the possibility of rethinking the space of thought characterising the eighteenth century. This is not to propose conclusive evidence or ascertain firm resolutions but to take aim at the incompatibility of landscape theory in general, and bring to light challenges that, like the chase, lead one further into the woods.

The notion of nature

The idea of a choreographed nature seems like a *contradictio in adjecto* unless the masterly invisible hand controlling and arranging plants, hillocks, lakes, rivers and the view on a summer afternoon would be an almighty and infinitely clever god orchestrating all from tiny little details such as flowers and branches, to the more grandiose elements like rising mountains, furiously flowing cataracts and endless, impenetrable forests. This suggests nature as a given, whether credited to a divinity or to biological evolution.

Generally speaking, nature appears at first to be a readily accessible term for which we may feel confident to possess a relatively precise definition. For example, nature is what is outside the urbanised areas; nature is green and organic; nature includes a wide range of different elements such as an uninhabited wilderness, the nearby-situated woods and the private paradise of the suburban garden. We might also consider that the concept of nature itself is rebellious, against interpretation and beyond comprehension. If nature is that which is untouched by human hands and the impact of civilisation, we have indeed very few, if any, areas in the world that might still be considered as "nature". What we normally term "nature" is much more likely what should be understood as the cultural or cultured landscape.

The so-called "natural garden" is a contradiction in terms. The notion of nature and the wish to conceptualise as well as deconstruct the term has been a vital part of an increasing number of discussions for the past decade. Choreographed nature refers to the elaborate and exquisite landscape gardens of the eighteenth century that were conceived as nature incarnate and the epitome of all the best of nature, but carefully contrived by a combination of ingenious gardeners and wealthy landowners. In fact these landscape gardens were not at all natural but a knowing and splendid staging of the environment according to principles of aesthetic ideals. Nature, in this instance, was a highly skilled construction originated by eighteenth-century men of taste.

An investigation between definitions of landscape and of picturesque gardens is equally mired in confusion. In 1792, William Gilpin, the British originator of the idea of the picturesque and the author of the influential publication *Three Essays: On Picturesque Beauty* specifically opted for "a distinction ... between such objects as are *beautiful*, and such as are *picturesque* – between those, which please the eye in their *natural state*; and those, which please from some qualities, capable of being *illustrated in painting*".[3] Furthermore, he elaborated on the importance of uniting the different parts into a whole, engaging the "picturesque eye" in an ongoing close examination of the various parts and bits.[4] In this process the form and composition is of great significance, but even more valuable are "atmosphere" and "various effects" derived from an embellished scene of nature, enabling what is termed "one comprehensible view".[5]

The notion of the "natural" was itself a cultural construction, and the outcome of the numerous and divergent interpretations that have been presented through the years is in itself a reflection of this condition. The following discussion of the process from concept by way of visualisation through to materialisation exposes the embedded ambiguity. The objective is a thorough discussion of the garden's status as manifesto and as an important media for aesthetic, as well as philosophical and political discussions. The eighteenth century was an epoch where the relationship between nature and culture underwent a dramatic change, in interaction with the ideological basis of the Enlightenment.

A natural garden

The epistolary novel *Julie, or the New Heloise: Letters of Two Lovers who Live in a Small Town at the Foot of the Alps* by the French Enlightenment philosopher Jean-Jacques Rousseau has become an extremely popular and repeatedly quoted source to the eighteenth-century concept of "natural" gardening and nature in general.[6] Only one specific letter, part 4, letter 11 to Milord Edouard, relates to this topic. Here Julie's Elysium, as she has named her private, secluded garden, hidden from the worldly sphere, is described in all details. The garden occupies an intriguing transitory zone, mediating between the private apartments of the mansion, the semi-public space of the landscape garden and the encircling scenery. The focus is on facilitating the passage from the home to the park, and as such the Elysium could be regarded as an exterior prolongation of the architecture carried through by other means.

The principal idea is the impression of a place untouched by human hands. It seems completely unspoiled, a masterpiece of nature. In spite of that, everything is masterly orchestrated, or as Julie more modestly expresses it in the book: "'It's true …true,' … she said, 'that nature has done everything, but under my direction, and there is nothing here which I have not ordered.'"[7] Furthermore art has been a vital ingredient. The tiny paths draw arabesque-like figures, and these traces contribute to a graceful balletic setting, where brooks alternate with running rivulets.

The managing of the layout includes everything from large-scale elements such as the controlling of the visual perception of the landscape outline, the staging of the garden as an exquisite space of experience, to the more detailed intervention where all sorts of plants and greenery have been treated like ornamental draping to enhance the overall impression of a combined work of art and nature. All together the elements of nature are consciously modified in order to form a highly elaborate setting, where the outdoor set pieces provide the framework for the sensitive stroll. Through movement we engage ourselves with the world.[8] In the case of Julie's Elysium, the knowledge of nature's transformable potentials makes it an important educational project as well. What appears as a first nature, an untouched virgin landscape, turns out to be a second or even third nature created by a virtuous and sensitive citizen as an important part of the project of cultivating future generations.[9]

The far-reaching influence of Rousseau's writings and philosophy in general in the late eighteenth century is visible in one of the major treatises on garden art, the five-volume *Theorie der Gartenkunst* by C.C.L. Hirschfeld, where Rousseau is a true hero fighting unremittingly against "*den falschen Geschmack in den Gärten*".[10] Following an extensive quotation from *Julie*, Hirschfeld emphasises that Elysium is "*ländlich, gleichsam vernachlässiget und doch reizend*". This fictional Elysium became a highly influential exemplar of the natural garden and a point of reference in unfolding arguments and debates on the subject. As a purely fictional construct it was on the one hand a product of an emerging sensibility already implemented in existing landscape parks while on the other it became decisive in the cementation of the same sensibility, raising it to an indispensable idiom.

Pavlovsk as encyclopedia of landscape design

Garden history is enriched with wonderful examples of great gardens created through the centuries, embedding ever-changing concepts and ideals, whether sylvan glades, arboreal landscapes, false perspectives, or formal parterres. It makes it quite difficult to pick out just one single garden that can serve as the quintessential representative.

Nature choreographed

It has already been pinpointed that the common denominator for gardens is the idea of a work in progress. Regarded as a work of art, the garden is a work of constant and rapid change. On a small scale the changing seasons contribute to important variations, from the blooming and promising spring to the fading flowers and the withering of autumn. On the larger scale, what is conceived as a grandiose layout eventually has to take off as a premature draft, waiting to develop through years and decades, waiting to ripen. It points to a unique dimension in garden art, setting it apart from, for instance, architecture, painting, sculpture and music. Gardens, by their very nature reliant on growth, require time to reach maturity and to fulfil design expectations. Any theoretical approach to landscape architecture and garden art must consider the temporal complexities of seasonal constraints and the atmospheric vicissitudes of the weather.

The Russian landscape garden Pavlovsk, situated 20 miles outside St Petersburg in Russia, provides a visually articulate case study. Pavlovsk offers an excellent example to illuminate the potentials and complexity of the landscape garden as an object for analytical inquiry. The area, initially described as thousands of acres of "woodlands, ploughed fields" and including "two villages with peasants" was the generous donation of Catherine the Great to her son Pavel Petrovich, after whom the park was named, and his German-born wife Maria Feodorovna on the occasion of their first-born son in 1777.[11] It soon transformed into an imperial residence with a well-formed castle and a stunning landscape park that evolved in the following decade. At Pavlovsk all the finest features of Western European park design, from decorative and ornamental elements to architectural structures, combined with specific components from the indigenous Russian landscape characterised by huge birch forests, winding rivers and brooks in an atmosphere imbued with lyrical melancholy. As such it is a unique blend of the well-known, familiar tradition and carefully selected influences from abroad that include classical landscape elements, pavilions inspired from ancient Greece and the Roman culture, and a picturesque landscape enhanced by the Slavianka river. In one area clumps of trees become wing screens dividing larger areas in successive numbers of compositions.

Pavlovsk has been termed "an encyclopedia of landscape architecture" and is indeed such.[12] For example, several monuments commemorate the loss of beloved ones, and other sentimental and romantic structures pay homage to bygone times in the shape of pastoral buildings that include a hermitage, an old chalet and the *ferme ornée* which is a perfectly functioning dairy with cows, sheep and a thoroughly rural exterior. Through the decades changing architects and gardeners have carried out Maria Feodorovna's master plan. In the first decade of the nineteenth century, painter and scenographer Pietro Gonzaga, one of those who, along with Vincenzo Brenna, Giacomo Quarenghi and Carlo Rossi among others, carried out the designs (albeit with different artistic ideals), addressed the beauty of the local countryside and incorporated it as a vital part of the composition. He enlarged the park in order to recreate the modest Nordic scenery in the shape of a flat terrain where a long "English" road found its way through the huge dense white birch forest where clumps of trees alternated with solitary plantings. Gonzaga combined

Figure 19.1
The Mausoleum of Tsar Pavel I covered in winter snow in the New Sylvia Region, 1801

Figure 19.2
The Pil Tower erected by Vincenzo Brenna in 1795

Nature choreographed

Figure 19.3
The sham ruin at Pavlovsk

Figure 19.4
View from the triple Lime Alley towards Pavlovsk Palace, designed by Scottish architect Charles Cameron in the early 1780s

nature and culture with various means that drew on aesthetics, principles of interior design and scenography. To him, vision was the predominant sense. In a influential treatise from 1807 with the captivating title *La Musique des yeux*, Gonzaga argues that if the most basic and primitive definition of music is the art of embellishing what is heard, then one might say that "*l'art d'embellir ce qu'on voit est la musique des yeux*".[13] An indispensable part of creating music for the eyes and mastering *le clavecin oculaire* is the right understanding of the potentials of the place, which means the ability to rightly orchestrate the various means.[14] Gonzaga goes on to suggest that "*le sage jardinier, dans la distribution de son plan, se propose donc une succession de scènes différentes, artistement préparées selon la qualité du projet et la disposition de son terrain*".[15] In addition to this advice to the wise gardener to create a succession of different views, Gonzaga painted *trompe l'oeil* frescoes of colonnades, imaginary staircases and statuary.

The park at Pavlovsk is a composite landscape, a bricolage place uniting the local and universal in which one finds everything that was currently in vogue in garden design. It presents a concentrate of the world, of nature and of culture. As a world within a world, the park presents different eras in different places within the grounds. Cultural codes from very different contexts seem reconciled in a masterly choreography that includes Italian antiquity, Russian national character, and elements from France and England all in a happy coexistence. Pavlovsk is a consciously created set-up, a grandiose interpretation of the world. Originally when the layout was carried out more than 200 years ago, it was an interpretation of something already existing, whether a Roman sculpture as *Apollo Belvedere*, an ostensible ruin, similar to the ones found in the landscape paintings by Hubert Robert, or an antique temple. In any discussion of eighteenth-century garden art, it is less important to clarify the translation of original and copy than to examine the processes by which these elements will be transformed, and the aesthetic as well as ideological context into which they will be translated. The landscape park is a highly sophisticated artefact and therefore the contextualisation is of outmost importance. The universe of garden art is complex and it is comparable to a Chinese puzzle of referentiality and inter-referentiality.

Perception of space in the landscape garden is often understood through the ambling or peripatetic walk in which the body's movement can be discursive, and one's thoughts and vision alternate between the contemplative observation and the quick glance. A recurrent theme in this connection is to what extent these gardens became decisive in shaping the modern gaze, sensibility and mentality. It is an important argument that a profound understanding of the complexity of eighteenth-century garden art will be a substantial and absolutely necessary component contributing to the upcoming modernity in as far as the garden achieves status as a manifesto where visions and ideals materialise. Landscape architecture and garden art are subordinated within the conditions of always being "in the making". This process of something "in the making" is a confrontation where ideas, concepts and visions meet and cope with the forces of nature, changing values of the cultural landscape, cultural factors in general and shifting habits. The starting point, where concepts are generated that consider the general layout of the garden, initiates a process which never entirely leads to a definite, everlasting result.

Pavlovsk reflecting historical changes

This discussion of Pavlovsk proffers a particularly interesting example of the consequences of this condition, since, to a quite unusual extent, Pavlovsk conforms to the agenda of the *great history* and political development in general. The owners of Pavlovsk through the nineteenth century were descendants of the extensive Romanov family with divergent approaches to the art of garden layout. However, even though small changes and improvements were carried out, the main ideas held sway.

When revolution came in 1917, Pavlovsk was a fairly intact testimony to the Russian imperial days of glory. The victory of Bolshevism proved to be the demise of the secure future for Pavlovsk as well as other imperial residences. The palaces were opened to the people; the parks functioned as military training camps. It was not until the end of the 1930s when Joseph Stalin conceded that the understanding of history and the fundamental codes of culture were an important asset of the Russian people and the communist classless society that an end was put to this vandalism and rather uncontrolled handling of the cultural heritage. The rough treatment of the communists was never the less unequalled to the Nazis' ravage and vandalising during the Second World War under the siege of Leningrad. When the defeated Germans left Russia in 1944, the palace was burned down to the ground, the forest decimated, the landscape completely destroyed and the plundering left barely anything, except for a few precious items that have been recently restored.

Our ability today to experience the heady magnificence of imperial grandeur at Pavlovsk is due to two circumstances. First, everything was systematically registered and mapped to the tiniest detail so that it was actually possible to reconstruct the original setting. And secondly the Russian people dedicated impressive and remarkable voluntary resources to carry out this reconstruction from the period right after the war until the late 1970s. The reconstruction in itself is full of paradoxes, and raises a lot of questions such as: Does it give additional meaning that the impoverished post-Second World War former USSR would allocate scarce resources to both restore and revive what might be considered a historic anachronism from its imperialist past? Other questions include: How might one ever aspire to recreate the "original" Pavlovsk? Would one start the reconstruction with 1780 plans and drawings or the nineteenth-century painter's depiction of the castle and grounds? Or, perhaps, the most recent registrations from just before the outbreak of the Second World War would provide the most authentic verisimilitude? And finally: What kinds of artificial reality will today's visitor experience, when travelling to the reconstructed Pavlovsk?

Apparently it is a landscape garden from the late eighteenth century, but even at that time it was a recreation of a kaleidoscopic imagery and the combined Arcadian visions from a bygone Italy and various landscape scenes from contemporary eighteenth-century gardens all over Europe. And in the end it is a late twentieth-century reconstruction, an image of the world and as such an interpretation of the life-world.

Throughout the years of the Pavlovsk's lifetime, changing agendas have thus been setting the scene. Consequently, Pavlovsk has been termed "an encyclopedia of landscape architecture, reflecting all the main trends in European eighteenth- and nineteenth-century garden design".[16] In fact, in the light of the events of the twentieth century described above, one could say that the Pavlovsk is encyclopedic even of later periods. Thus, the story of Pavlovsk has been written and rewritten continuously for more than 200 years. As such Pavlovsk is an artefact constantly in the making, continuously materialising shifting and transient ideas and values. Part of this complexity is what W.J.T. Mitchell clarifies in his fourth *Theses on Landscape*.

> Landscape is a natural scene mediated by culture. It is both a represented and presented space, both a signifier and a signified, both a frame and what a frame contains, both a real place and its simulacrum, both a package and the commodity inside the package.[17]

Mitchell's point is that it is not important or even possible to decide in favour of any of these sharply opposed binaries. What is important is to confront and address the lag between the two. The landscape garden is interesting as an object of study because it is either a "real place" or a "simulacrum". To some extent one can claim that what makes it look like a simulacrum, at the same time is what makes it a real place. Or, to return where we began the discussion with Foucault's *heterotopia*, that which is one place and all places simultaneously.

Conclusion

The possibility to rethink the critical potentials of garden art and landscape design raises more questions than it answers and depends on what kinds of conclusion one wishes to put forward. The complexity of the object of study calls for an interdisciplinary approach that will be open and non-reductive, and further maintains the tension between the universal and the particular. In the interpretative gesture it is indeed tempting to reduce the artefacts to a common denominator, an essential core, or a generative first principle, which makes things easy to handle, fitting in perfectly, and reinforcing a given idiom. Due to this intermediary status that garden art and landscape design can claim to hold, garden art may be regarded as a critical medium capable of thinking with and against the epistemological space of a given period and changing agendas as each era proffers its own scene change. It could be a privileged space, an empirical order and "a middle region", as Foucault terms it, when he pins down the position between the already encoded eye and reflexive knowledge. Following Foucault, who writes about the ability any culture possesses to access "the pure experience of order and its modes of being", garden and landscape design offer the possibility of unfolding between "the ordering codes and reflections upon order itself".[18] The rather eclectic approach one finds in the mode of relating to tradition and inspirational sources in late eighteenth-century garden art could even argue for pointing to a certain resemblance to the postmodern idiom.

Notes

1. Simon Swaffield (ed.), *Theory in Landscape Architecture: A Reader*, Philadelphia, PA: University of Pennsylvania Press, 2002.
2. Michel Foucault, *The Order of Things: An Archaeology of the Human Sciences*, London: Routledge, 1996, p. xiv.
3. William Gilpin, *Three Essays: On Picturesque Beauty; On Picturesque Travel; and on Sketching Landscape – to which is Added a Poem, on Landscape Painting*, London: R. Blamire, 1792, p. 3.
4. *Ibid.* 44.
5. *Ibid.*
6. Jean-Jacques Rousseau, *La Nouvelle Heloîse – Julie or the New Eloise*, Judith H. McDowell (trans.), Philadelphia, PA: University of Pennsylvania Press, 1761/1987.
7. *Ibid.* 305.
8. Michel Conan (ed.) *Landscape Design and the Experience of Motion*, Washington DC: Dumbarton Oaks Research Library and Collection, 2003.
9. John Dixon Hunt, *Greater Perfection: the Practice of Garden Theory*, London: Thames & Hudson, 2000.
10. C.C.L. Hirschfeld, *Theorie der Gartenkunst vol. I–V*, Hildesheim, NY: Georg Olms Verlag, band 1: 129, (1779–1785) 1973.
11. Nikolai Tretiakov, *Pavlovsk: Palace and Park, Art-Palace*, St Petersburg: Art-Palace, copyright Pavlovsk State Museum-Reserve, 2005, p. 6.
12. Anatoli Kuchumov, *Pavlovsk: Palace and Park,* Leningrad: Aurora Art Publishers, 1975, p. 277.
13. Pietro Gonzaga, *La Musique des yeux et l'optique théatral,* St Petersburg: Chez A. Pluchart, imprimeur du département des affaires étrangéres, p. 12, 1807.
14. *Ibid.* 57.
15. *Ibid.* 45.
16. Suzanne Massie, *Pavlovsk: The Life of a Russian Palace*, Boston, IL, Toronto and London: Little, Brown and Company, p. 82, 1990.
17. W.J.T. Mitchell (ed.), *Landscape and Power*, Chicago, IL and London: The University of Chicago Press, p. 5, 2002.
18. Foucault, p. xxi.

Chapter 20

The birth of modernity out of the spirit of music?

Henry van de Velde and the Nietzsche Archive

Ole W. Fischer

In early January 1889, the German philosopher Friedrich Nietzsche, a professor of philology in early retirement from the University of Basel, sent several obscure letters to his friends and colleagues from Torino signed "Dionysos," "Nietzsche Caesar" or "The Crucified."[1] When his alarmed friend Franz Overbeck from Basel arrived in Torino on 8 January he found Nietzsche out of his mind, crying, screaming and playing the piano naked, so he decided to take him back to Basel's asylum, the first station of Nietzsche's eleven-and-a-half-year twilight in madness. Yet in the aftermath of Nietzsche's breakdown, his manuscripts, as well as his letters and parts of his library remained in Torino and scattered at various places of his unsteady life between cheap hotels in Italian cities and alpine villages, and visiting friends in Basel and family in Naumburg. With the ebbing hope for Nietzsche's mental recovery, the question arose of what to do with his literary remains, especially since the last months of his conscious life in fall 1888 were extraordinarily productive.

Nietzsche's mother was overburdened with the home care for her paralyzed philosopher son. Although as a Lutheran pastor's widow, she was repelled by his radical writings, she gave her acknowledgement to both Franz Overbeck and Heinrich Köselitz (alias Peter Gast), Nietzsche's former student and secretary, to act as literary executors. Together with Nietzsche's publisher they started the edition of his collected writings, but things changed rapidly in 1893 when Friedrich's younger sister Elisabeth Förster entered the scene. After the final return from a failed "Germanic" colony experiment in Paraguay, where she lost her wealth, her mission and her husband Bernhardt Förster, who was a Wagnerian and notorious anti-Semite, she immediately took over the self-declared legacy as representative of Nietzsche's interests, and seized his literary remains.² She collected his manuscripts, struggled with the publisher for the proof sheets with his annotations, asked all his correspondence partners for a return or copy of his letters, and collected his library and private papers. In 1894 Elisabeth Förster-Nietzsche, the name she used from then on, was able to open the Nietzsche Archive on the first floor of her mother's house in the small city of Naumburg – with the doomed philosopher upstairs. Soon her literary circles, afternoon teas, piano soirées and other social activities interfered with the care for the sick brother, and Förster-Nietzsche chose to move to the small residency in Weimar, to participate in the social life of the Grand Duke's court, and profit from the glorious heritage of Friedrich Schiller and Johann Wolfgang Goethe, the emblems of German classic literature. In spring of 1896, after staying provisionally in a rented apartment, a grant by Meta von Salis, a writer friend of Nietzsche, provided Förster-Nietzsche with a villa on top of the hills overlooking Weimar. The death of their mother had made it necessary to reunify once again both the archive and the care for the mad philosopher brother under the same roof: Nietzsche himself was transported at night in a special train cabin from Naumburg to Weimar.

In the meantime Förster-Nietzsche had twice started an edition of Nietzsche's writings. Although she was in charge of copyrights of the published works and had managed to collect almost all the literary remains, she several times disassociated from the editors she had engaged. First she fired Heinrich Köselitz, who had been working as an editor since 1893, then she hired and fired Fritz Kögel, Rudolf Steiner (later the founder of anthroposophy), as well as Ernst and August Honeffer, within a few years. Finally, in 1898, it was Köselitz again, the only one able to read Nietzsche's cryptic handwriting, who came back and helped her to start the edition project of the 20-volume *Complete Works*, which was not finished before 1913.³ This *Großoktaveausgabe* (large octavo edition) together with the following soft-cover edition, Nietzsche's *Choices* and *Collected Letters* were an overwhelming success, soon translated into French, English and many other languages, and became the basis of Nietzsche's continuous fame until today.⁴ However, the Nietzsche Archive remained a private institution, or more precisely a one-woman property, which brought unfortunate side effects to the publishing policy. Förster-Nietzsche held back Nietzsche's finished, but unpublished autobiographical book *Ecce homo*, manipulated several of his letters, and, with the uncritical help of her editors, compiled his so-called masterpiece *Will to Power* from fragments of various sketchbooks.⁵ In addition, she vindicated her own image of her brother with a series of biographies.⁶ This met criticism right from the beginning. Former friends of Nietzsche like Overbeck, former editors and employees of Förster-Nietzsche, as well as other Nietzsche experts protested, but Förster-Nietzsche recognized the strong faith in her by almost all public intellectuals of her time. And she still held in her hand the "tragic philosopher," the "fallen eagle" and most important "piece" in her collection, whom she used to show to "special guests" of the archive. With the example of Cosima as high priestess of the Richard Wagner cult in Bayreuth in mind, she cultivated her role of devoted sister, wise woman and hostess of a cultural circle in Weimar, and, with the help of the patron Count Harry Kessler, the archive soon turned into a center of early modern avant-gardes.

Förster-Nietzsche understood the importance of art and media in modern society (as well as the new laws on copyright) and so she monopolized the production of Nietzsche portraits, sculptures and photographs by various artists. In fact, she made use only of photographs of her brother from the time before his breakdown, whereas she preferred paintings, etchings and sculptures of the sick philosopher with a prophetic attitude. Finally, with the help of Kessler, she succeeded in finding with Hans Olde, Edvard Munch and Max Klinger the corresponding artists who were able to fulfill her claim and represent Nietzsche between intimate martyr and heroic prophet. In addition, Förster-Nietzsche handed out various pieces and fragments of Nietzsche's writings to several of the new art and literature magazines, which were emerging around the turn of the century in Berlin, Vienna, Darmstadt, Frankfurt, Munich and Leipzig. These publications helped her to connect the philosophy of Nietzsche with the new aesthetic movement of Art Nouveau, Secession and Jugendstil, and to raise the demand for Nietzsche's books. The increase of income through donations and royalties made it possible for Förster-Nietzsche not only to pay for transcribing, correcting and editing Nietzsche's works, but also to enjoy bourgeois comfort. As early as 1898 she commissioned alterations to the archival villa "Silberblick" (gleam of silver), which de facto still belonged to Meta von Salis, and Förster-Nietzsche's actions led directly to the break-up of this friendship. Yet with Nietzsche's death in August 1900 she was in need of a new attraction for the archive and so she followed the advice of Count Kessler to opt for an interior design by the style reformer Henry van de Velde.

The Flemish painter, autodidact designer and architect van de Velde was recognized as an early enthusiastic follower of the "philosopher with the hammer," and after he had his breakthrough at the art exhibition of Dresden in 1896 as "inventor" of German Art Nouveau, his star began to shine over the same circles that were interested in Nietzsche's "New Man." In spring 1901 he gave a lecture series about the theoretic foundation of the "New Style" in the famous literary salon of Cornelia Richter in Berlin, and Count Kessler, at one of the following soirées in his apartment designed by the artist, introduced van de Velde to Förster-Nietzsche. She in turn invited van de Velde and Kessler on a *pèlerinage* to Nietzsche's tomb in the small village of Röcken on 25 August 1901, the first anniversary of the philosopher's death. Van de Velde, who, with the bankruptcy of his arts-and-crafts workshops in Berlin and Brussels in 1900, had given up the idea of a new guild society and was in search of a new field of activity, became immediately interested in her plan to reform the applied arts production in the Grand Duchy of Weimar. Förster-Nietzsche had it in mind to reanimate the idea of a cultural Weimar movement: after the "golden age" with the poets Schiller and Goethe, and the silver age with the composer and virtuoso Franz Liszt, she imagined a "New Weimar" of arts, architecture and life reform with the help of van de Velde under the banner of Nietzsche's philosophy. To reinforce her and Count Kessler's diplomatic maneuvers for van de Velde's appointment at court, she charged the artist with the alteration of the Villa Silberblick in fall 1901. At the same time this building transformation was an expression of the new autonomy Förster-Nietzsche had gained by taking over the ownership of the archival villa that year. Van de Velde, on the other hand, who had found in the early 1890s in Nietzsche's radical writings the source for his own artistic "mission," welcomed the commission to remodel the archive as a chance to combine his interests in aesthetic reform with a homage to "his" philosopher.

Even if the shifts of perspective in Nietzsche's philosophy are legendary, a close reading can identify some continuous threads of thought over the subsequent periods of his writings. One of these red ribbons is Nietzsche's distrust of language, morality and convention. Here he alternates between laying bare "lie" as the origin of language (instead of communication and understanding) and acknowledging the stabilizing effect of "false beliefs" (such as Christian morals) for the evolution of society on the one hand versus the fundamental rejection of his contemporary culture as "decadence" on the other. Nietzsche's account of the fatal crisis of society, language, tradition and art was applied by his reader van de Velde to his own situation as a socially marginalized avant-garde painter and transformed into a fundamental critique to expose bourgeois culture and reject academic artistic production. Instead van de Velde started theorizing and propagated a renewal of the fine arts under the leadership of applied arts and architecture that would finally give birth to the "New Style." Yet this radical change should not be limited to art and aesthetics alone, since van de Velde thinks art in close resonance with the societal relations. Both the philosopher and the artist consider the disintegration of the arts an analogy to the disintegration of the modern individual into various social roles that lead to imitation, staging and play. In contrast they propose the unity of art and life, and imagine the "New Man" as builder-creator "beyond good and evil" (that is, free from traditional moral beliefs), and as authentic, whole "great ego," for whom they refer to historic figures such as Napoleon or Goethe.

Charged with high ambitions of legitimating his hopes for a new beginning in Weimar by giving an example of his "New Style" as well as expressing personal gratefulness to the philosopher who made him change his life and take up the "mission" of life reform, the design of the Nietzsche Archive holds a special place in the oeuvre of van de Velde. Whereas his previous architectonic and interior works are dominated by graphic curvature and extravagant dynamics, the archive breathes an alternative spirit of intellectual solemnity, integral harmony and introvert concentration. And whereas he started for most of his designs with an ornamental line as a gestural act – an archaic force channeled out of immediate emotion before the rationalizing effect of language – he made use of an alternative artistic strategy for the Nietzsche Archive – what he called "ornamental transcription." The term is used in music theory to describe the notation of music, respectively the rewriting of a piece for another instrument or arrangement. It is especially the second meaning that sheds light on van de Velde's intentions to relate architectural form to philosophic thought rendered after the model of music, even more so, as he uses alternatively the term of programmatic art.

As a former music critic and married into a family of virtuosi, van de Velde was well aware of this aesthetic concept of referring to external thoughts of philosophy or literature with internal means of the discipline, as it was formulated in late-romantic music theory by Richard Wagner and Franz Liszt, and he transferred those ideas to applied art, architecture, furniture and book design. Programmatic art was meant to disarm the latent intellectual distrust put forward against music (and architecture) by the aesthetic systems of the German idealistic philosophy of Kant, Hegel and Schelling, who preferred the conceptualized artwork of poetry and philosophy. They disregarded music (and architecture) as meaningless entertainment or emotional expression, and therefore as inferior art. According to Wagner, music is able to refer to an external "program" by the title of the work, an explanatory theoretic text of the composer (from where the term program derives), and a significant way to structure the abstract material into themes or so-called leitmotifs. Wagner goes on to explain that Beethoven consciously transgressed the canonic symphonic form with the vocal finale of his Symphony No. 9 (after Schiller's *Ode to Joy*) in order to transcend and express highest emotion: the celebration of a liberated mankind. Beethoven's finale was for Wagner the rebirth of the *Gesamtkunstwerk* of ancient Greek tragedy. Van de Velde adopted this idea of the synthesis of arts interpenetrating all aspects of life. But even more relevant for van de Velde's aesthetic thought was the rejection of mimicry and imitation in the concept of programmatic art, providing a possibility to inform an abstract object with philosophic meaning beyond application of symbolic ornament or classical tectonic language. Nietzsche, who had reflected on Wagner, Schopenhauer and the metaphysics of music in his early writings, proposed yet another important motive of non-figural representation: he suggested the identity of internal and external worlds, a resemblance of character and whole of the post-Christian thinker with his built environment, which reformulates the pre-Socratic idea of *physis* as an organic unity of spirit and matter, of surface and content, of inside and outside, or what Nietzsche called the "architecture for the perceptive."

> Architecture for the perceptive. – There is and probably will be a need to perceive what our great cities lack above all: still, wide extensive places for reflection; places with tall, spacious, lengthy colonnades for inclement or unduly sunny weather where no traffic noise or street cries are able to penetrate, and where a refined sensibility would forbid even a priest to pray aloud: buildings and locations that express as a whole the sublimity of stepping aside to take thought for oneself. The time has passed when ... the vita contemplativa primarily had to be a vita religiosa ....We want to have ourselves translated into stones and plants; we want to have ourselves to stroll in, when we take a turn into those porticoes and gardens.[7]

For the Nietzsche Archive van de Velde operated with a series of artistic manipulations that can be read as "programmatic": he improved the unsatisfactory entrance of the house by adding a street-facing portico to the simple cubic building. To mark

Henry van de Velde and the Nietzsche Archive

Figure 20.1
Villa Silberblick, Weimar, 1898 (before van de Velde's alterations)

Figure 20.2
Villa Silberblick/Nietzsche Archive, 1904 (after van de Velde's alterations)

its status as a public institution, he labeled the new entrance with the inscription Nietzsche Archive carved in stone in broad Roman letters. This gesture did not quite correspond to the status of a private villa, but had to be understood in the context of programmatic art as the "title" of this work. For building the new portico van de Velde continued to use the materiality of brick and stucco of the existing structure, but rather than resembling the logic of wall and opening of the neo-Renaissance original, the new street facade is a compositional play of surfaces and proportions. This anthropomorphic positioning of openings can be directly connected to Nietzsche's idea of physiognomic expression, as in the "architecture for the perceptive." The excessive height of the entrance door in dark oak serves as part of this geometric frame, but for the approaching guest it offers another enigma: instead of a door handle there is a set of sculptural bronze handholds with labyrinth ornamentation. This might reflect on the unclear status of the house between shrine and

Figure 20.3
Henry van de Velde. New street facade (portico) and entrance situation of the Nietzsche Archive, Weimar, 1904

villa, between literary archive and last domicile of the philosopher, but at the same time it structured the proportion of power of inside and outside: the arriving guest had to request access. In addition, van de Velde noted in an earlier version of his memoirs to the chapter of "The *Nietzsche Archive* and the New Weimar," that he had intended to give the archive an appearance "more solemn and monumental like a *Schatzkammer* (treasure chamber)"[8] – the leitmotif of this work.

Once inside, after a flight of stairs, there is a dark entrance hall. A crystalline lamp over the doorway illuminates cloakrooms with series of brass coat hooks.[9] These additional spaces work as a joint between the capitalist chaos of the city outside and the synthesis of art inside, constructing a new society of "Nietzscheans." A few steps to the right a double door opened to the "treasure" of the archive: the library with "his" books and manuscripts. This oblong room, a merging of two smaller rooms, has a rather low ceiling for its size. Since van de Velde could not easily change the height within an existing house, he chose a repetitive vertical structure as organic "ribs" to arrange the walls and virtually elevate and carry the white plafond. These planks hold the shelves for books, but integrated openings with movable window grilles prohibit unwanted visitors. The color palette of this room ranges from natural red beech, to fraise-colored plush and intensive-red curtains, heightened by white stucco and brass details, only contrasted by the grayish-blue carpet. The room as a whole was meant to invoke the atmosphere of an alpenglow of Nietzsche's *Zarathustra*. Van de Velde put another "title" reference inside his work, this time in the form of the initial of the philosopher's name: a brass N in a circle is embedded above the tiled stove. Nietzsche himself, apart from his books on

Henry van de Velde and the Nietzsche Archive

the shelves and his manuscripts in the cupboards, is "present" in a life-size marble stele by Max Klinger. The only object in the room not designed by van de Velde is this stele, which rests on a platform against a surface of colored glass illuminated by evening light.

But why did van de Velde deliberately blur the status of the main room between private salon and sacred temple, literary archive and intimate library? The answer might be found in the program of the "New Weimar" and its direct rivalry with the cult of Goethe, manifest in the conversion of Goethe's house in Weimar to a national museum in 1885, as well as the new Goethe–Schiller Archive, built in 1896. The Nietzsche Archive had to stand against a comparison with the palazzo of the thinker, poet and minister with its exquisite classical interiors and artwork. Goethe had brought back the idea of the humanist *studiolo* from his Italian journey of 1786 to 1787 and remodeled his house into a personal microcosm: the succession of salons, dining hall, study chamber, scientific collection, garden and library

Figure 20.4
Henry van de Velde. Nietzsche Archive, Weimar, interior view from the west

Figure 20.5
Henry van de Velde. Nietzsche Archive, Weimar, interior view from the entrance (to the east, note the tiled stove with the monumental N and the stele by Max Klinger at the back)

were read as an ideal portrait of the educated bourgeois. Van de Velde's strategy of staging a mood of authenticity, a plausible, but retroactive *studiolo* for the dead philosopher was extraordinary successful. For Nietzsche, who had never consciously understood that he had vegetated for four years in Weimar, van de Velde created a physiologic resemblance of architecture and philosophy, and constructing an organic atmosphere for "the perceptive" with a synthetic work of art in his new style: the interiors of the Nietzsche Archive (and van de Velde's book illustrations) soon became synonymous with "Nietzsche design," providing evidence of the modernity and superiority of "his" philosopher.

After the Second World War, the Nietzsche Archive was closed down because of its associations with the Nazi regime, manifest in an unfinished neoclassicist Nietzsche Memorial Hall next to the archival villa started in 1938. Nietzsche's manuscripts and books were confiscated by the East German authorities, together with the literary remains of Förster-Nietzsche, and transferred to the socialist predecessor of the Weimar Classics Foundation, the *Nationale Forschungs- und Gedenkstätte (der Deutschen Klassik)* in Weimar. Since Georg Lukács had denounced Nietzsche's philosophy as proto-fascism, there was almost no opportunity for serious research on archival stocks of Nietzsche in East Germany.[10] The building of the archival building was hidden by the new owners, the inscription destroyed, and the villa modified and reused as a seminar building and guest house of the socialist National Research and Memorial Place in Weimar.[11]

In the 1960s, the Italian philosopher Giorgio Colli and the Italian philologist Mazzino Montinari started their project of a new critical edition of Nietzsche's works. An ideological re-evaluation of Art Nouveau and early modernism in the 1960s and 1970s opened the frame for a renovation of the archive building as well as its interior, which was begun in 1984, already five years before the fall of the Berlin Wall, and remained unfinished until 1991 in the reunited Germany.

Today the Nietzsche Archive is a museum of the national Weimar Classics Foundation and therefore it is open to the public, but the manuscripts are stored in the Goethe–Schiller Archive. Friedrich Nietzsche's as well as his sister's books belong to the Anna Amalia Library of the same Weimar Classics Foundation, and are only accessible for institutional research. After all, and with the irony of history, the private archive of the philosopher's sister was united with its national rival, and Nietzsche's original writings even became part of the world's cultural heritage, but not in the sense imagined: the Nietzsche Archive is an archive with empty shelves.[12]

Notes

1 Friedrich Nietzsche, *Sämtliche Briefe. Kritische Studienausgabe. Band 8*, ed. Giorgi Colli/Mazzino Montinari, Berlin and New York: Walter de Gruyter, 1986; 2003, p. 567–79.
2 H.F. Peters, *Zarathustra's Sister. The Case of Elisabeth and Friedrich Nietzsche*, New York: Crown, 1977; Carol Diethe, *Nietzsche's Sister and the Will to Power. A Biography of Elisabeth Förster-Nietzsche*, Urbana, IL: University of Illinois Press, 2003.

3 Friedrich Nietzsche, *Nietzsche's Werke* (*Großoktav-Ausgabe*), Leipzig: Naumann (1899–1909); Leipzig: Kröner (1909–); in fact as late as 1926 the last volume – an index of the previous 19 volumes of works and fragments – was published.
4 The *Weimarer Nietzsche Bibliography* of 2002 counted over 22,000 publications of Nietzsche's works or that deal with Nietzsche (up to the deadline in 1998)!
5 The book *Ecce homo*, finished by Nietzsche in 1888, remained unpublished until 1908, since Förster-Nietzsche was worried about several offensive paragraphs that would strengthen voices that interpreted Nietzsche's work as "pathologic"; in addition there were several paragraphs where Nietzsche explicitly criticized his sister and mother.
6 Elisabeth Förster-Nietzsche, *Das Leben Friedrich Nietzsche's. 2 Bände*, Leipzig: Naumann 1895–1904; Elisabeth Förster-Nietzsche, *Der junge Nietzsche*, Leipzig: Kröner, 1912; Elisabeth Förster-Nietzsche, *Der einsame Nietzsche*, Leipzig: Kröner 1914; Elisabeth Förster-Nietzsche, *Wagner und Nietzsche zur Zeit ihrer Freundschaft*, München: Müller 1915; Elisabeth Förster-Nietzsche, *Friedrich Nietzsche und die Frauen seiner Zeit*, München: C.H. Beck, 1935; the inaccuracy of the edition of the *Nietzsche Archive* was obvious, but there was no better edition until 1967 with the critical edition of Giorgi Colli and Mazzino Montinari: *Friedrich Nietzsche Werke. Kritische Gesamtausgabe*, still, the final transcription and edition of the handwritten fragments remained unfinished up to today – 109 years after the death of Nietzsche.
7 Friedrich Nietzsche, *Die fröhliche Wissenschaft, Buch 4, § 280, Kritische Studienausgabe* [KSA]. Band 3, ed. Giorgi Colli and Mazzino Montinari, Berlin and New York: Walter de Gruyter, 1988, p. 524–25.
8 Henry van de Velde, *Manuscript*, in Bibliothèque Royal, Brussels, *Archives Henry van de Velde*, FS X 1-2, p. 414: "*solennel et monumental d'une »Schatzkammer«*," citation after Henry van de Velde, *Récit de ma vie. II*, ed. Anne van Loo and Fabrice van de Kerckhove, Brussels and Paris: Versa/Flammarion, 1995, p. 155, annotation 1.
9 Crystal as a metaphoric motif in Nietzsche's *Zarathustra* influenced expressionist architects like Paul Scheerbart and Bruno Taut.
10 Georg Lukács, *Der deutsche Faschismus und Nietzsche*, Paris: C.A.L.P.O. 1945; Georg Lukács, *Die Zerstörung der Vernunft*, Berlin: Aufbau-Verlag, 1954.
11 Affected by this alteration was the second floor of the Nietzsche Archive with the private chambers of Elisabeth Förster-Nietzsche and the death room of Friedrich Nietzsche, which were destroyed at that time, but which were not touched by van de Velde's alteration in 1902–1903; some of the furniture remained in the depots of the Goethe National Museum in Weimar.
12 The Goethe–Schiller Archive in Weimar is part of the UNESCO program Memory of the World since 2001.

Select bibliography

Alberti, L.B., *L'Art d'Edifier*, Texte traduit du latin, présenté et annoté par Pierre Caye et Françoise Choay, Ouvrage traduit et publié avec le concours du Centre National du livre, Paris: Editions du Seuil, 2004.
Alberti, L.B., *On Painting and on Sculpture*, The Latin texts of De Pictura and De Statua, Edited with translations, introduction and notes by Cecil Grayson, London: Phaidon Press Limited, 1972.
Alberti, L.B., *On the Art of Building in Ten Books*, translated by Joseph Rykwert, Neil Leach and Robert Tavernor, Cambridge, MA, and London: The MIT Press, 1988.
Alberti, Romano, *Origine e Progresso dell'Academia del Disegno*, Pavia: Pietro Bartoli, 1604.
Allen, Stanley, "Libeskind's Practice of Laughter: An Introduction by Stan Allen", in *Assemblage* 1990: 12, pp. 20–25
Alpharani, Tiberii, *De Basilicae Vaticanae antiquissima et nova structura* [1582], introduction by Michele Cerrati, Rome: Tipografia Poliglotta Vaticana, 1914.
Aquinas, Thomas, *The Basic Writings of Saint Thomas Aquinas*, Anton Pegis (ed.), New York: Random House, 1945.
Aristotle, *Metaphysics*, Louise Loomis (ed.), New York: Walter Black, 1943.
Arroyo, Ciriaco Morón, *The Humanities in the Age of Technology*, Washington, DC: The Catholic University of America Press, 2002.
Bachelard, Gaston, *La poétique de l'espace*, Paris: Presses Universitaires de France, 1957.
Bacon, Francis, "Of the Wisdom of the Ancients" (1609), in Joseph Devey (ed.), *The Moral and Historical Works of Lord Bacon*, London: George Bell and Sons, 1888, pp. 200–268.
Banham, Reyner, *Theory and Design in the First Machine Age*, London: Architectural Press, 1960.
Baron, H., *In Search of Florentine Civic Humanism: Essays on the Transition from Medieval to Modern Thought*, Vol. I, Princeton, NJ: Princeton University Press, 1988.
Basalla, George, *The Evolution of Technology*, Cambridge: Cambridge University Press, 1988.
Baxandall, M., *Giotto and the Orators, Humanist Observers of Painting in Italy and the Discovery of Pictorial Composition, 1350–1450*, London and Oxford: Clarendon Press, 1971.
Benevolo, L., *Storia dell'architettura moderna*, Bari: Laterza, 1960.
Benjamin, Walter, "The Work of Art in the Age of Mechanical Reproduction", *Illuminations*, trans. Harry Zohn, New York: Schocken, 1968.
Benjamin, W.," Theses on the Philosophy of History" (1940), in Arendt, H. (ed) *Illuminations: Essays and Reflections*, New York: Harcourt Brace Jovanovich, 1968, pp. 253–264.
Bhabha, Homi K., *The Location of Culture*, London: Routledge, 1994.
Bloom, Harold, *The Anxiety of Influence*, New York: Oxford University Press, 1973.
Boccaccio, G., *Il Decameron ... di nuouo ristampato, e riscontrato in Firenze con testi antichi, & alla sua vera lezione ridotto dal cavalier Lionardo Salviati ...* Appresso Pietro Maria Bertano, Venice, 1638.

Bibliography

Borromeo, Carlo, *Instructiones fabricate et suppellectilies ecclesiasticae, Parte Seconda, Regulae et instructions de nitore et munditia ecclesiarum, atriorum, sacrorum locorum et suppellectilis ecclesiasticae* (1577), edited by Z. Grosselli, Milano: Università Cattolica, 1983.

Borst, Arno, *The Ordering of Time: From the Ancient Computus to the Modern Computer*, Cambridge: Polity Press, 1993.

Braudel, Fernand, *The Mediterranean World in the Age of Philip II*, Sian Reynolds (trans.), Glasgow: William Collins, 1972.

Breton, André, *Je vois j'imagine: poèmes-objets*, Paris: Gallimard, 1991.

Breton, André, *Arcanum 17*, translated by Zack Rogow, Los Angeles: Green Integer, 2004.

Breton, André, *Surrealism and Painting*, Boston: New York: MFA Publishers, 2002.

Brooks, Allen H., *Le Corbusier's Formative Years: Charles-Edouard Jeanneret at La Chaux-de-Fonds*, Chicago, IL: University of Chicago Press, 1997.

Bunschoten, Raoul, "Wor(l)ds of Daniel Libeskind, Daniel Libeskind: Theatrum Mundi: 20 February – 23 March 1985. Three Lessons in Architecture 1985 Venice Biennale", in *AA Files* 1985: 10, pp. 79–88.

Burckhardt, J., *The Civilization of the Renaissance in Italy, an Essay*, London: Phaidon Press Ltd, 1965.

Carl, Peter, "The Godless Temple, 'organon of the infinite'", *Journal of Architecture*, Vol 10, Number 1, 2005, pp. 63–90.

Carroll, L., *Alice's Adventures in Wonderland*, New York: Appleton and Co, 1866.

Cassirer, E., Kristeller, P.O., Randall Jr., J.H. (eds.), *The Renaissance Philosophy of Man*, Chicago, IL, and London: University of Chicago Press, 1948.

Cassirer, E., *The Individual and the Cosmos in Renaissance Philosophy*, Translated with an introduction by Mario Domandi, Mineola, NY: Dover Publications Inc., 2000.

Cassirer, E., *The Philosophy of Symbolic Forms, Volume One: Language*, translated by Ralph Manheim, preface and introduction by Charles W. Hendel, New Haven, CT, and London: Yale University Press, 1955.

Chiles, James R., *Inviting Disaster: Lessons from the Edge of Technology*, New York: HarperCollins, 2001.

Choay, F., *The Rule and the Model, On the Theory of Architecture and Urbanism*, Edited by Denise Bratton, Cambridge, MA, London: The MIT Press, 1997.

Cicero, *De Finibus Bonorum et Malorum*, with an English translation by H. Rackham, Cambridge, MA: Harvard University Press, 1914.

Clément, Catherine, *The Lives and Legends of Jacques Lacan*, New York: Columbia University Press, 1983.

Colomina, Beatriz, *Privacy and Publicity: Modern Architecture as Mass Media*, Cambridge, MA: The MIT Press, 1994.

Conan, M. (ed.), *Landscape Design and the Experience of Motion*, Washington DC: Dumbarton Oaks Research Library and Collection, 2003.

Cowan, Ruth Schwartz, *A Social History of American Technology*, New York and Oxford: Oxford University Press, 1997.

Cusa, N. of, "Idiota de Mente", in *Nicholas Of Cusa, On Wisdom and Knowledge*, Hopkins, J. (ed.), Minneapolis, MN: The Arthur J. Banning Press, 1996, pp. 528–601.

Cusanus, Nicolaus, "La perception de la profondeur: Alhazen, Berkeley et Merleau-Ponty", *Oriens-Occidens* 2004: 5, pp. 171–84.

Cusanus, Nicolaus, "*ON KAI KHÔRA*: Situating Heidegger between the *Sophist* and the *Timaeus*", *Studia Phaenomenologica* 2004: 4, pp. 73–98.

de Blaauw, Sible, *Cultus et decor. Liturgia e architettura nella Roma tardoantica e medievale: Basilica Salvatoris, Sanctae Mariae, Sancti Petri*, Vatican City: Biblioteca Apostolica Vaticana, 1994.

Deleuze, Giles, *Woran erkennt man den Strukturalismus?* Berlin: Merve Verlag, 1992.

De Saussure, Ferdinand, *Course in General Linguistics*, Roy Harris (trans.), London: Duckworth, 1983.

Dickens, Charles, *Great Expectations*, London: Penguin Classic, 1966.

Dolar, Mladen, *A Voice and Nothing More*, Cambridge, MA: The MIT Press, 2006.

Bibliography

Eberhart, Mark E., *Why Things Break: Understanding the World by the Way it Comes Apart*, New York: Three Rivers Press, 2003.

Eck, C. van, "The Structure of De Re Aedificatoria Reconsidered", *Journal of the Society of Architectural Historians*, 1998: 57, 3, pp. 280–297.

Eisenman, Peter, "Post-Functionalism", in *Oppositions* 1976: 6, unpaginated.

El-Bizri, Nader, "In Defense of the Sovereignty of Philosophy: al-Baghdâdî's Critique of Ibn al-Haytham's Geometrisation of Place", *Arabic Sciences and Philosophy* 2007: 17, pp. 57–80.

Evelyn, John, *The Diary*, Ernest Rhys (ed.), 2 vols., London: J. M. Dent and Sons Ltd., 1936.

FitzGerald, Michael, *Picasso. The Artist's Studio*, Exhibition catalogue, New Haven, CT: Yale University Press/Wadsworth Atheneum Museum of Art, 2001.

Florensky, Pavel, *Iconostasis*, New York: St Vladimir's Seminary Press, 1997.

Floryan, M., *Gardens of the Tsars: A Study of the Aesthetics, Semantics and Uses of Late 18th Century Russian Gardens*, Århus: Aarhus University Press, 1996.

Förster-Nietzsche, Elisabeth, *Friedrich Nietzsche und die Frauen seiner Zeit*, München: C.H. Beck, 1935.

Forty, Adrian, *Words and Buildings*, London: Thames and Hudson, 2000.

Foucault, M., *The Order of Things – An Archaeology of the Human Sciences*, London: Routledge, 1996.

Foucault, Michel, *Death and the Labyrinth: The World of Raymond Roussel*, London, New York: Continuum, 2004.

Francia, Ennio, *1506–1606: Storia della Costruzione del Nuovo San Pietro*, Roma: De Luca Editore, 1977.

Frascari, Marco, *Monsters of Architecture*, Lanham, Maryland: Rowman & Littlefield Publishers, Inc. 1991.

Freud, Sigmund, *Civilization and its Discontents*, in *The Standard Edition of the Complete Psychological Works*, vol. 21, James Strachey (trans. and ed.), (1961 rpt.) London: Vintage, The Hogarth Press and the Institute for Psychoanalysis, 2001, pp. 64–145.

Frommel, S., *Sebastiano Serlio: Architetto*, Milan: Electa, 1998.

Galassi Paluzzi, Carlo, *La basilica di S. Pietro*, Bologna: Cappelli Editore, 1975.

Gallop, Jane, "Lacan's 'Mirror Stage': Where to Begin?", *SubStance* 11, 4, Issue 37–38, A special issue from the Center for Twentieth Century Studies, 1983: pp. 118–128.

Giedion, S., *Space, Time and Architecture: The Growth of a New Tradition* (1940), Cambridge: Harvard University Press, 1967.

Giedion, Siegfried, *Mechanisation Takes Command: A Contribution to Anonymous History*, New York: Oxford University Press, 1948.

Gies, Francis, and Joseph Gies, *Cathedral, Forge, and Waterwheel: Technology and Invention in the Middle Ages*, New York: HarperCollins, 1994.

Gilpin, W. (1792) *Three Essays: On Picturesque Beauty; On Picturesque Travel; and on Sketching Landscape – to which is Added a Poem, on Landscape Painting:* London.

Gonzague, P. G., *La Musique des Yeux et L'optique Théatral*, St Petersbourg, 1807.

Gracq, Julien, *42 rue Fontaine. L'Atelier d'André Breton*, Paris: Adam Biro, 2003.

Grayson, C., "The Humanism of Alberti", *Italian Studies*, 1957, XVI, pp. 37–56.

Hanafi, Zakiya, *The Monster in the Machine: Magic, Medicine, and the Marvelous in the Time of the Scientific Revolution*, Durham, NC: Duke University Press, 2000.

Harries, Karsten, "Modernity's Bad Conscience", *AA Files*, no. 10, Autumn 1985.

Harries, K., *Infinity and Perspective*, Cambridge, MA, London: The MIT Press, 2001.

Hays, Michael K., *Modernism and the post-Humanist Subject*, Cambridge, MA, and London: The MIT Press, 1992.

Heidegger, Martin, *Sein und Zeit, Gesamtausgabe Band 2*, Frankfurt am Main: Vittorio Klostermann, 1977.

Hirschfeld, C.C.L., *Theorie der Gartenkunst Vol. I–V* (1779–1785), Hildesheim and New York: Georg Olms Verlag, 1973.

Hopkins, J., "Nicholas of Cusa", in *Dictionary of the Middle Ages*, Vol. 9, Strayer, J.R. (ed.), New York: Charles Scribner's Sons, 1987, pp. 122–125

Hunt, J.D., *Greater Perfection – The Practice of Garden Theory*, London: Thames & Hudson, 2000.

Bibliography

Jarzombek, Mark, "Readymade Traces in the Sand: The Sphinx, the Cimera, and the Other Discontents in the Practice of Theory", in *Assemblage* 1992: 19, pp. 73–95.
Jenkins, Ian, "Ideas of Antiquity: Classical and Other Ancient Civilizations in the Age of the Enlightenment", in Kim Sloan and Andrew Burnett (eds.), *Enlightenment: Discovering the World in the Eighteenth Century*, London: The British Museum Press, 2004, pp. 168–177.
Jouffroy, Alain, "La collection André Breton", *L'Oeil*, no. 10, October 1955, pp. 32–39.
Joyce, James, *Finnegans Wake*, New York: Viking Press, 1967.
Kachur, Lewis, *Displaying the Marvelous*, Cambridge, MA: The MIT, 2001.
Kaufmann, E., *Architecture in the Age of Reason, Baroque and post-Baroque in England, Italy, and France*, New York: Dover Publications, Inc., 1968.
Kelly, Kevin, *Out of Control: The New Biology of Machines*, London: Fourth Estate, 1994.
Klibansky, Raymond, Erwin Panofsky, and Fritz Saxl, *Saturn and Melancholy, Studies in the History of Natural Philosophy, Religion, and Art*, New York: Basic Books, 1964.
Koyré, Alexandre, *Etudes d'histoire de la pensée scientifique*, Paris: Gallimard, 1973.
Kühn, Paul, *Das Nietzsche-Archiv in Weimar*, Darmstadt: Alexander Koch, [1904].
Lacan, Jacques, *Écrits*, Paris: Editions de Seuil, 1966.
Lampugnani, Vittorio M., "Daniel Libeskind: Between Method, Idea and Desire", in *Domus* 1991: 731, pp. 17–28.
Le Corbusier, *Toward an Architecture*, intro. Jean-Louis Cohen, trans. J. Goodman, Los Angeles, CA: Getty Research Institute, 2007.
Le Corbusier, *Vers une architecture*, Paris: les Éditions G. Crès, 1923.
Le Corbusier, *The Decorative Art of Today* [based on *L'Art décoratif d'aujourd'hui* published in 1925 by G. Crès et cie, Paris], Cambridge, MA: The MIT Press, 1987.
Lévi-Strauss, Claude, *The Savage Mind*, London: Weidenfeld and Nicolson, 1966.
Libeskind, Daniel, "Three Lessons in Architecture: Architecture Intermundium", in Daniel Libeskind, *Radix-Matrix, Architecture and Writings*, Munich and New York: Prestel, 1997, pp. 64–69.
Libeskind, Daniel, "Three Lessons in Architecture: The Machines", in Daniel Libeskind, *The Space of Encounter*, New York: Universe Publishing, 2000, pp. 180–194.
Lindberg, David C., "Lines of Influence in the 13th century Optics: Bacon, Witelo, and Pecham", in *Speculum*, Vol. 46, no. 1, January 1971, pp. 66–83.
Lindenbaum, Peter, *Changing Landscapes: Anti-Pastoral Sentiment in the English Renaissance*, Atlanta, GA and London: University of Georgia Press, 1986.
Lippmann, W., *A Preface to Morals*, New York: The Macmillan Company, 1929.
Lyotard, Jean-François, *The Postmodern Condition: A Report on Knowledge*, Minneapolis, MN: University of Minnesota Press, 1984.
Massie, Suzanne, *Pavlovsk: The Life of a Russian Palace*, Boston, MA, Toronto and London: Little, Brown and Company, 1990.
Matussek, Peter, "The Renaissance of the Theatre of Memory", in *Janus* 2001: 8, pp. 4–8.
Mazzotta, Giuseppe, *Cosmopoiesis: The Renaissance Experiment*, Toronto: University of Toronto Press, 2001.
McGonagall, William, *The Tay Bridge Disaster and other Poetic Gems*, Washington, DC: Orchises Press, 2000.
McGuire, W. (ed), *The Freud/Jung Letters: The Correspondence between Sigmund Freud and C.G. Jung*, London: Penguin, 1991.
Millon, Henry and Vittorio Magnano Lampugnani, *The Renaissance from Brunelleschi to Michelangelo, The Representation of Architecture*, New York: Rizzoli, 1994.
Mitchell, W.J.T. (ed.), *Landscape and Power*, second edition, Chicago, IL, and London: University of Chicago Press, 2002.
Motycka Weston, D., "'Communicating Vessels': André Breton and his Atelier, Home and Personal Museum in Paris," *Architectural Theory Review*, Vol. II, No. 2, 2006, 101–128.
Motycka Weston, Dagmar, "Le Corbusier and the Restorative Fragment at the Swiss Pavilion", in Mari Hvattum and Christian Hermansen (eds) *Tracing Modernity: Manifestations of the Modern in Architecture and the City*, London: Routledge, 2004.
Nauert Jr, Charles G., *Humanism and the Culture of Renaissance Europe*, Cambridge: Cambridge University Press, 1995.

Bibliography

Nazîf, Mustafâ, *Geometry and Dioptrics in Classical Islam*, London: *al-Furqân* Islamic Heritage Foundation, 2005.

Nazîf, Mustafâ, *Les mathématiques infinitésimales* IV, London: *al-Furqân* Islamic Heritage Foundation, 2002.

Nietzsche, Friedrich, *Sämtliche Briefe. Kritische Studienausgabe*, edited by Giorgi Colli and Mazzino Montinari, Berlin and New York: Walter de Gruyter, 2003.

Nietzsche, Friedrich, *Sämtliche Werke. Kritische Studienausgabe* [KSA], edited by Giorgi Colli and Mazzino Montinari, Berlin and New York: Walter de Gruyter, 1988.

Onians, J., "Alberti and ΦΙΛΑΡΕΤΗ, A Study in their Sources", *Journal of the Warburg and Courtauld Institutes*, 1971: XXXIV, pp. 96–114.

Orlan, "Intervention", Chapter Nineteen, *The End(s) of Performance*, in Peggy Phelan and Jill Lane (eds), New York: New York University Press, 1997.

Pacey, Arnold, *The Maze of Ingenuity: Ideas and Idealism in the Development of Technology*, (1974 2nd edn), Cambridge, MA and London: The MIT Press, 1992.

Panofsky, E., *Idea, Contribution à l'Histoire du Concept de l'Ancienne Théorie de l'Art*, Traduit de l' Allemand par H. Joly, Préface de J. Molino, Paris: Editions Gallimard, 1989.

Panofsky, Erwin, *Idea, A Concept in Art Theory*, Columbia, SC: University of South Carolina Press, 1968.

Peden, Margie, et al. (eds.), *World Report on Road Traffic Injury Prevention: Summary*, Geneva, World Health Organization, 2004.

Perec, Georges, *Life: A User's Manual*, David Bellos (trans.), London: Godine, 2000.

Pérez-Gómez, A. and Pelletier, L., *Architectural Representation and the Perspective Hinge*, Cambridge, MA, and London: The MIT Press, 1997.

Pérez-Gómez, Alberto, "Abstraction in Modern Architecture: Some Reflections in Parallel to Gnosticism and Hermeneutics", *VIA, Journal of the Graduate School of Fine Arts, University of Pennsylvania*, Vol. 9, 1988, pp. 71–83.

Peters, H.F., *Zarathustra's Sister: The case of Elisabeth and Friedrich Nietzsche*, New York: Crown, 1977.

Pinelli, Antonio, *The Basilica of Saint Peter's in the Vatican*, Modena: Edizioni Panini, 2000.

Piotrowski, Andrzej, "The Spectacle of Architectural Discourses", *Architectural Theory Review*, 2008: 13, 2: 130–144.

Plato, *Euthyphro. Apology. Crito. Phaedo. Phaedrus*, translated by H.N. Fowler, Cambridge, MA: Harvard University Press, 1999.

Plotinus, *The Six Enneads*, Stephen MacKenna and B. S. Page (trans.), Chicago, IL: Encyclopedia Brittanica, Inc., 1952.

Polizzotti, Mark. *Revolution of the Mind: The Life of André Breton*, London: Bloomsbury, 1995.

Preziosi, Donald, ed. *The Art of Art History: A Critical Anthology*, Oxford and New York: Oxford University Press, 1998.

Rhodes, Richard, *Arsenals of Folly: The Making of the Nuclear Arms Race*, London: Simon and Schuster, 2008.

Rodenbeck, John, "Travellers from an Antique Land: Shelley's Inspiration for 'Ozymandias'", *Alif: Journal of Comparative Poetics* 24, 2004, pp. 121–148.

Rossi, A., *Scritti scelti sull'architettura e la città. 1956–1972*, Milan: CittàStudi, 1975.

Rossi, A., *The Architecture of the City*, Cambridge MA and London: The MIT Press, 1982.

Rossi, A., *Selected Writings and Projects*, London and Dublin: Gandon Editions, 1983.

Rousseau, J.-J., *La nouvelle Heloîse – Julie or the New Eloise* (1761), translated by Judith H. McDowell, University Park, PA: The Pennsylvania State University Press, 1987.

Ruskin, John, *The Seven Lamps of Architecture*, [based on the issue published in 1880 by George Allen, Sunnyside, Orpington, Kent; first published in 1849], New York: Dover, 1989.

Rykwert, J., *On Adam's House in Paradise: The Idea of the Primitive Hut in Architectural History*, 2nd edn, Cambridge, MA: The MIT Press, 1981.

Rykwert, J., *The Dancing Column: On Order in Architecture*, Cambridge, MA: MIT Press, 1996.

Sabra, Abdelhamid I. (trans.), *Ibn al-Haytham's Optics, Books I–III: On Direct Vision*, London: Warburg Institute, 1989.

Said, Edward W., *Culture and Imperialism*, New York: Knopf, 1993.

Bibliography

Sawday, Jonathan, *Engines of the Imagination: Renaissance Culture and the Rise of the Machine*, London: Routledge, 2007.

Semper, Gottfried, *The Four Elements of Architecture and Other Writings*, ed. and trans. Harry F. Mallgrave and Wolfgang Herrmann, Cambridge: Cambridge University Press, 1989.

Shakespeare, William, *The Complete Works*, Stanley Wells and Gary Taylor (ed.), Oxford: Clarendon Press, 1988.

Shelley, Mary, *The Last Man*, (1826), Pamela Bickley (ed.), Ware: Wordsworth Editions, 2004.

Silvan, P., "Le origini della Pianta di Tiberio Alfarano", *Atti della Pontificia Romana Accademia di Archeologia* (Rendiconti), 1992: LXII 1989–1990, pp. 3–25.

Silverman, Debora, "A Fin-de-Siècle Interior and the Psyche: The Soul Box of Dr Jean-Martin Charcot", *Daidalos* 28, 1988, pp. 24–31.

Slingerland, E.G., *What Science Offers the Humanities*, Cambridge: Cambridge University Press, 2008.

Spengler, Oswald, *Man and Technics: A Contribution to a Philosophy of Life*, Honolulu, Hawaii: University Press of the Pacific, 2002.

Steiner, George, *The Idea of Europe, Introductory Essay by Rob Riemen*. Nexus Library IV, 2005.

Swaffield, S. (ed.), *Theory in Landscape Architecture – A reader*, Philadelphia, PA: University of Pennsylvania Press. 2002.

Tafuri, M., *Architecture and Utopia. Design and Capitalist Development*, Cambridge MA: The MIT Press, 1976.

Tafuri, M., *Theories and History of Architecture*, London: Granada. Cambridge, MA and London: The MIT Press, 1980.

Tafuri, M., *Interpreting the Renaissance. Princes, Cities, Architects*, D. Sherer (trans.), New Haven, CT: Yale University Press, 2006.

Tainter, Joseph, *The Collapse of Complex Societies*, Cambridge: Cambridge University Press, 1988.

Taylor, A.J.P., *War by Timetable: How the First World War Began*, London: Macdonald and Co., 1969.

Temple, N., "Architecture and the Humanist Tradition", in *Four Faces – The Dynamics of Architectural Knowledge, The 20th EAAE Conference*, Stockholm, Helsinki, 2003, pp. 159–161.

Tinniswood, Adrian, *By Permission of Heaven: The Story of the Great Fire of London*, London: Pimlico, 2004.

Tzonis, Alexander and Liane Lefaivre, *Classical Architecture. The Poetics of Order*, Cambridge, MA: The MIT Press, l986.

van de Velde, Henry, *Récit de ma vie. II. 1900–1917*, edited by Anne van Loo and Fabrice van de Kerckhove, Brussels and Paris: Versa/Flammarion, 1995.

Vaughan, Diane, *The Challenger Launch Decision: Risky Technology, Culture, and Deviance at NASA*, Chicago, IL, and London: University of Chicago Press, 1996.

Verene, Donald Phillip, *Knowledge of Things Human and Divine: Vico's* New Science *and* Finnegans Wake, New Haven, CT: Yale University Press, 2003.

Vesely, D., *Architecture in the Age of Divided Representation, The Question of Creativity in the Shadow of Production*, Cambridge, MA, and London: The MIT Press, 2004.

Vesely, Dalibor, "Architecture and the Ambiguity of the Fragment", in R. Middleton (ed.) *The Idea of the City*, London: The Architectural Association, 1996.

Vesely, Dalibor, *Architecture and Continuity: Kentish Town Projects 1978–81, Diploma Unit 1*, London: Architectural Association, 1982.

Vico, Giambattista, *The Autobiography of Giambattista Vico*, Max Harold Fisch and, Thomas Goddard Bergin (trans.), Ithaca, NY: Cornell University Press, 1944.

Vico, Giambattista, *The New Science of Giambattista Vico*, Thomas Goddard Bergin and Max Harold Fisch (trans.), Ithaca: Cornell University Press, 1984.

Vidler, Anthony, "The Building in Pain: The Body and Architecture in Postmodern Culture", in *AA Files* 1991: 19, pp. 3–10.

Virilio, Paul, *The Original Accident*, Julie Rose (trans.), Cambridge: Polity Press, 2007.

Vitruvius, *The Ten Books on Architecture*, New York: Dover Publications, 1960.

von Moos, Stanislaus, *Le Corbusier: Elements of a Synthesis*, Cambridge, MA: The MIT Press, 1979.

Wagner, David, *The Seven Liberal Arts in the Middle Ages*, Bloomington, IN: Indiana University Press, 1986.
Weintraub, Linda, *Art on the Edge and Over: Searching for Art's Meaning in Contemporary Society*, Litchfield: Art Insights, 1996.
Yates, Francis A., *The Art of Memory*, London: Routledge and Kegan Paul, 1966.
Young, J.Z., *An Introduction to the Study of Man*, Oxford: Clarendon Press, 1971.
Zevi, B., *Storia dell'architettura moderna*, Turin: Einaudi, 1950.
Zuccari, Federico, *L'Idea de' Pittori, Scultori ed Architetti*, Torino, 1607, in Detlef Heikamp (ed.), *Scritti d'Arte di Federico Zuccaro*, Firenze: Leo S. Olschki Editore, 1941.

Index

9/11 (events of) 58

Acropolis 112
Abbott and Costello 172, 182n.1
aesthetic(s) 10, 233, 241; engineer's 73
Alberti, Leon Battista xvii, 120, 126, 131, 136–146, 166–7; *De Pictura* 126;
De Re Aedificatoria 120, 136–146; Santa Maria Novella 165
Alberti, Romano 169; *Origine et progresso dell' Academia del Disegno* 161
Alfarano, Tiberio 120, 147–51, 155
Alhazen (al-Hasan ibn al-Haytham) 126–7, 129–31; *Optics* (*De Aspectibus, Perspectiva* or *Prospettiva*) 126–7, 130–41
Alice in Wonderland (Lewis Carroll) 107
Angkor Wat 47
ANY 214
Apollonius of Perga 129–30
Aquinas, Thomas 166; *Summa Theologica* 166
Aragon, Louis 203
Arcadia 106, 113, 235
Arcanum 17 211
Archimedes 129

architectural, drawing 119, 157; knowledge 137; history 215; language 215; project 214–5, 218–9, 223; thought 55
Arequipa, Incan settlement of 111
Aristotle 4, 12, 36, 38, 41, 44, 68, 127, 130, 163–8;
Arab Aristotelian thought xix; Lyceum 23; *Metaphysics* 34, 120, 160, 163–4; *Physics* 130; *Rhetoric* 169
Arizona 66
Arroyo, Ciriaco Moron xiv–xvii, 55
Art Nouveau 239, 245
artificial intelligence 194; computer 8
artist 73, 95, 120, 160, 163–8, 196, 202–3, 207, 212, 240; artist's studio xii, 11, 187, 204–5, 208
aura 13–14
Avenarius, Richard 194
Aymonino, Carlo 222

Bachelard, Gaston 123
Bacon, Francis 55, 60; *De sapientia veterum* 59
Bacon, Roger; *Opus maius*; *De multiplicatione specierum* 126
Bali 47

Index

Ballard, J. G., *The Drowned World* 66; *The Drought* 66; *The Burning World* 66
Balzac, Honoré de 203
Barbaras 195
Baron, Hans 142
Basalla, George 59–60
Basel 126
Basra 126
Baudrillard, Jean 41
Baxandall, Michael 142
Beckham, David 7
Beaufret, Jean 22
beaux-arts 29
Beethoven 241
Belting, Hans, 155
Bellori, Pietro, *Vite de' pittori, scultori et architetti moderni* 218
Benevolo, Leonardo, *Storia dell'architettura moderna* 218
Benjamin, Walter 12–14, 56, 93, 99, 224
Benz, Karl 57
Berkeley, George 132
Bernini, Gian Lorenzo 45; *Ecstasy of St Theresa* 168
Bhabha, Homi K. 71
Bilbao 47, 112
Biraghi, Marco 217, 224
Blondel, Jean François 46
Bloom, Harold 179
Boccaccio, Giovanni, *Decameron* 107
Bohemianism 47
Bolshevik revolution; Bolshevism 74, 234
Bonnier de la Mosson, Joseph 209
Borges, Jorge Luis 92, 96, 98
Borromeo, Carlo 154, 180
Borromini, Francesco and Pietro 161, 166, 167; Santi Luca e Martina 120, 161; San Carlo alle Quattro Fontane 120, 161, 165, 167–8
Bouch Thomas 63
Bramante, Donato 150–51, 153

Braque, George 204
Braudel, Fernand 92, 94–5, 97, 99
Brasilia 47
Brecht, Bertolt 99
Brenna, Vincenzo 230
Breton, André 187–8, 201–213; his 'wall' assemblage 209–212; *Souvenir of Earthly Paradise* 212
Brown, Lancelot 'Capability' 46
Brunelleschi, Filippo 126
Bruni, Leonardo 142; *De Interpretatione Recta* 142
Brussels 239
Bufalini, Leonardo 154
Butler, Samuel, *Erewhon* 64

camera obscura 128
Cameron, Charles 232
Camillo, Giulio 83–4; *L'Idea del teatro, see also* Memory Theatre 84
capitalism 72
Carruthers, Mary 157
Casabella 214
Cassirer, Ernst 143
Castells, Manuel 47
Challenger, see also space shuttle 60
Chambers, William 46
Childish, Billy 93
Chiles, James 59
de Chirico, Giorgio, *Child's Game* 208
Choay, Françoise 137
Christ 45, 51, 154, 156–7; Church of 150
Christianity 45
Cicero 182n.4; *De Finibus Bonorum et Malorum* 143; *de Oratore* 163
city 17, 43, 47, 50, 76, 85, 104, 107, 112, 137, 139, 142, 212, 217, 219–223
civic order 124
Claro, Elisa 209, 211
Classical Greece 24, 50, 74, 23, 33, 37–8, 44

Clément, Catherine 179
Cohen, Jean-Louis 76
Cold War 63
Colli, Giorgio 245
Colomina, Beatriz 76
colonisation 49
Columbus, Christopher 45
consumerism 7, 49, 79, 202
Cortona, Pietro da, *Trattato della Pittura e Scultura* 120, 161
cosmos 139, 161
Cowan, Ruth Schwartz 61
Crusoe, Robinson (Daniel Defoe) 105
Crystal Palace 57
Cuban missile crisis 63
curiosities, cabinet of (*Wunderkammer*) 208; room of (*Schatzkammer*) 47, 243

Dada 51, 202
Daedalus 59
Dada 202
Dal Co, Francesco 43; *History of Modern Architecture* (see also Tafuri) 43
Dante Alighieri 7, 211
Darwin, Charles 7–8, 74; *Origin of Species* 60
decorum 46
Degottex, Jean; *Black Pollen* 209
Delage 1921 Grand-Sport car 77
Deleuze, Gilles 56, 92, 95, 97
Descartes, René 17, 131; Cartesian 10–11, 33; *res extensa* 132, 161; *res cogitans* 132, 161, 167
determinism 56, 73
Dickens, Charles 92; *Great Expectations* 91
Dilthey, Wilhelm 193–5
disegno 119–20, 144, 160–71
Driscoll, Bridget 57
Doomsday Machine 63
Duke of Savoy 45
Dupérac, Ètienne 151–52

durée 94
dwelling 3, 8, 19–20, 22, 25, 30–2, 38–40, 51, 56, 105, 107, 111

Eck, Caroline van 137
Edison, Thomas 59
Eden 106, 184
Edsall, Arthur 57
Edison, Thomas 59
Egypt 44; Cairo 126
eidos 155, 160, 165
Eisenman, Peter 56, 86, 217
Eluard, Paul 203
England 46, 74, 226; Brighton Pavilion 46; Croyden 57; Essex 92; Hastings 95; Kent 92; Lincoln xviii; Medway 92; Milton Keynes 92; Rochester 91; Southend 91
Enlightenment 9–10, 17–18, 20, 33, 36, 41, 47, 50, 66, 116, 121, 193, 226, 228
Erlach, Fischer von 45
Ernst, Max 203
L'Esprit nouveau 75
Estienne, Robert 155; *Thesaurus linguae latinae*
ethics xvi, xviii, 3–4, 16–27, 33–4, 38, 105, 116, 124, 138–43, 195, 203, 212, 226, responsibility of architecture 114, 142
Euclid 127, 129, 133; *Data* 131; *Elements* 131
Euclidean solids, *see also* geometry 76
Eurydice 180
Eurocentric attitude 74
Europe 45, 48, 66
Evelyn, John 62
everydayness 40, 188, 189–90, 199, 202
experimental method 126, 133
extimité (the extimate) 172–3, 177

fantasy 64, 68, 107, 179, 181, 183,
Farnsworth House 99–100

Index

Ficino, Marsilio, *Theologica Platonica* 169; *Sopra lo amore* 161, 166
First World War 64
Florentine Platonic Academy 166
Forster, E. M. 65
Foucault, Michel 235–6; *The Order of Things* (*Les Mots et les choses*) 86, 227
Forty, Adrian 87
France 47, 74, 226
Frank, Pat, *Alas, Babylon* 66
Frankel, Maria 179
Freud, Sigmund 65, 181, 203, *Civilisation and its Discontents* 65; uncanny (*Unheimlich*) 121, 173, 177, 181
French café 199
French Revolution 50
Frommel, Sabine 104
functionalism 86–7

Gadamer, Hans-Georg 29, 34, 189, 193, 195; *Truth and Method* 30, 193
Gallop, Jane 173, 176
Garden(s) 106–116, 188; of Eden 66, 105, 184; Chinese 46; design 226, 230, 233, 235–6; Kew 46
Gehry, Frank 47
genius loci 92
geometry 119, 130–41, 166; Euclidean 74, 76, 131
Gerard of Cremona 126
Germany 18, 74, 226; Basel 126, 237; Bayreuth 238; Berlin 16–18, 23–4, 239, 245; Darmstadt 9, 239; Dresden 239; Frankfurt 239; Leipzig 239; Naumburg 237–8; Munich 239; Röcken 239; New Weimar 188, 244; Weimar 238–9
Gesamtkunstwerk 241
Ghiberti, Lorenzo 131; *Commentario terzo* 126, 131
Ghirardo, Diane 217

Giacometti, Alberto, *Suspended Ball* 211
Gibbon, Edward, *History of the Decline and Fall of the Roman Empire* 66
Giedion, Siegfried, *Space, Time and Architecture* 218
Gilpin, William, *Three Essays: On Picturesque Beauty* 228
Ginzburg, Carlo 147
God 51, 62, 138, 160, 167, 184
Goethe, Johann Wolfgang von 7, 188, 194, 238–9, 244
Golgotha 156
Goncourt brothers 203
Gonzaga, Pietro 230; *La Musique des Yeux* 233
Goodman, John 76
Great Fire of London 62
Great Britain 226
Greco-Arabic 125
Gress, David 50
Grimm, Jacob 20
Grosseteste, Robert 133n.4
Guarini, Guarino 45–6

Hades 179–80
Hall, Sir Peter, *London 2000* 91
Harries, Karsten 30
Harvey, David 47
Hays, Michael K. 85
Hebrews 4, 51
Heidegger, Martin 9, 19–22, 32, 38–41, 50, 193, 195; *Being and Time* 20, 132; being-in-the-world 4, 21–22, 39, 132; *Bildung* 11, 32–3, 193; Dasein 21–22, 40; Heideggerian 4, 20; *Question Concerning Technology* 37
Hegel, G. W. F. 241
Heisenberg, Werner 196
Helmholtz, Hermann 194
hermeneutics xix, 195
Heraclitus 38–40
Hermes 177, 180, 184n.17,19

Index

Hirschfeld, C. C. L., *Theorie der Gartenkunst* 229
historiography xix
historicism 194
history, movement of 31; philosophy of 97; of architecture 125, 188; of the present 188, 215–6; of science 126; project of 215, 219, 223
Hitchcock, Alfred 96
Hitler, Adolf 96
Hobbes, Thomas 55, 67; *Leviathan* 66
Hodges, William 44
Hohokam 66
Homer 7, 116, 177, 183n.6; Odysseus 182n.1
Honeffer, August and Ernst 238
hubris 63
Hughes, Thomas P. 60
Hugo, Victor 201
humanism 38, 120, 142, 160; anti-humanist 41, 86; civic 142; *studia humanitatis* xvi-xvii, xix, 119–20, 128
Hume, David 18
Humpty Dumpty 178
Hunt, John Dixon 46
Husserl, Edmund 194–5, 201; lived world (life-world, *Lebenswelt*) 188, 194, 196, 201, 235
Huysmans, J. K., *A rebours* 203

illusion 26, 84, 100, 128, 176
imaginative universal (*universale fantastico*) 173
Indus Valley 66
Industrial Revolution 66
interior design 233
Ireland 45
Ise Shrine 105, 114
Israel *see also* Hebrews 18, 44
Italy 100, 137, 142, 235; Turin 45–6; Palmanova 85; Rome *see* Rome; Venice *see* Venice

Jefferson, Thomas 114, 116
Jeffries, Richard, *After London* 67
Jerusalem 16–19, 23–4, 104, 106
Joyce, James 43, 176; *Finnegans Wake* 182; *Ulysses* 181
Jugendstil 239
Julius II (Pope) 149–51
Jung, Carl 222

Kant, Immanuel 10–12, 14, 36, 41, 132, 241; *Critique of Historical Reason* 194; *Critique of Pure Reason* 132, 194
Keegan, John 64
Kent, William 46
Kepler, Johannes 126, 190
Kessler, Harry Count 239
Khwârizmî 130
Al-Kindî 133n.4
King Louis XIV 45
Klein bottles 180
Klinger, Meta von Salis Max 239, 244
knowledge 194–5; architectural 137; epistemology 73, 119, 133, 141, 163, 227; technical 55
Köselitz, Heinrich (alias Peter Gast) 237–8
Krier, Leon 87
Kubrick, Stanley 55, 64; *2001: A Space Odyssey* 64; *Clockwork Orange* 96; *Dr Strangelove or: How I Learned to Stop Worrying and Love the Bomb* 63–4
Kundera, Milan, 3, 12

labyrinth 59–60, 112, 116n.6, 167, 174, 180–81, 242
Lacan, Jacques 121, 172–185; mirror stage 121, 173, 176–7, 179–81
Lacis, Asja 99
Lagueux, Maurice 16
landscape 46, 226, 227–8
landscape architecture xix, 227, 230, 233, 235

Index

Lautrémont, Comte de 205
Lawrence, D. H. 65
Ledoux, Claude Nicolas 10, 12
Le Corbusier 11, 47, 55–6, 71–80, 212;
   Vers une architecture 55–6, 72,
   74–6, 79
Lechte, John 95
Lefaivre, Liane 12
Leibniz, Gottfried 50, 131
Leonidov, Leonid 190
Lessing, Doris, *The Memoirs of a*
   Survivor 66
Lévi-Strauss, Claude 94, 204
Libeskind, Daniel 56, 81–90, 190, 217
lineamentis 120, 136, 138, 138, 142–3,
   155
Liszt, Franz 239, 241
locus classicus 55
Lodoli, Carlo 184
logic 23, 191
London 57, 91–2; Docklands 47;
   Greenwich 91, 95; Hackney 95;
   Upper Norwood 57
Lomazzo, Gian Paolo, *Trattato dell-arte*
   della pittura 167
Lukás, Georg 245
Lyotard, Jean-François 22, 71

Mackenzie, Donald 59
Macrobius, *Commentary on the Dream*
   of Scipio 182
macrocosm 165, 209
Marion, Jean-Luc 195
Marquesas Islands 211
Marvell, Andrew 67
Marx, Leo 61, 66
Marx, Karl 59
mathematics 101, 124–5, 127, 129–30,
   132, 191, 194; Archimedean-
   Apollonian tradition of 126
Matisse, Henri 204
McGonagall, William 62–3
Mediterranean 74
Medusa 175

Melusina 211
memory x, xix, 65, 69, 75, 128, 156,
   158, 173, 175, 190, 196, 201
Memory Theatre 83
Merleau-Ponty, Maurice xviii, 195
Merton, Robert K. 58
Mesopotamia 66
Metafisica 175–7
Michelangelo 74, 78, 150–52, 156–7;
   Porta Pia 161
microcosm 165, 209
Middle Ages 50, 157
mi-dire (half speech) 179, 179, 183n.15
Miller, William M., *A Canticle for*
   Leibowitz 66
Militizia, Francesco 46
millennium 49
Milton, John 67
Mirandola, Pico della 45
Miró, Joan, *Head* 209, 211
Mitchell, W. J. T., *Theses on*
   Landscape 235
Möbius band 180
modernism 9, 29, 37, 46, 50, 72, 86–7,
   220, 245; crisis of 215–6
modern(ist) architecture 5, 43, 86
modernity xx-xxi, 32–3, 43–51, 65, 86,
   125, 188, 215, 233, 245
Monroe, Marilyn 96
mons delectus 174
Montaigne, Michel de, *Essays* 66
Montesquieu, Baron de, Charles-Louis
   de Secondat 51
Montinari, Mazzino 245
Moos, Stanislaus von 76
Moreau, Gustave 204
Moriston, Wesley 22
Morrison, William Percy 57
Munch, Edvard 237

nature 176, 226–236
Nazi 46, 234, 245
Neolithic 58
Neoplatonism xix, 127, 160–171

neo-Pythagorean 127
New Dispensation 51
New Guinea 209, 211
New Testament 63, 108
New Tribalism 49
Newton, Isaac 135n.28, 191–92
Nicholas of Cusa 120, 141, 187; *Idiota de Mente* 138
Nietzsche, Friedrich Wilhelm 95, 188, 237, 242; Bernhardt Förster (husband of Elizabeth); *Choices* 238; *Collected Letters* 238; *Complete Works* 238; *Ecce homo* 238; Elizabeth Förster-Nietzsche 188, 237–9; Nietzsche Archive 188, 237–246; Silberblick 239–40; *Will to Power* 238
Noah's Ark 111
North America 48, 111, 112
North British Railway Company 63

Old Testament 56, 66, 104–5, 106, 110; Abraham 51; Cain 51; Moses 44
Olde, Hans 239
optics 119, 123–35, 179
Orlan, Saint 10–11
Orpheus 180, 182n.4
Ortega y Gasset, José 8
Ottonelli, Domenico 161
Overbeck, Franz 237–8
Oxford and Cambridge Review 65
Oxford English Dictionary (OED) 16, 68

Paestum 77
Palestinians 19
Paraguay 237
Paris 45–6, 203, 205, 207, 209; Comedie de 207; *Gallerie des Beaux Arts* 211; Jardin des Plantes 209; Montmartre 205; Mouline Rouge 208; Pigalle 205
Parthenon 77
Patocka, Jan 195

Paul V (Pope) 149
Peckham, John, *Perspectiva communis* 126
pedagogy 107
Pelacani da Parma, Biagio, *Quaestionis perspectivae* 126
Pepin sisters 179
Perec, Georges 56, 92–3, *Life: A User's Manual* 93, 96–7
Pérez-Gómez, Alberto xvii, xix
Persepolis 44
Persians 44
perspectiva 125–6, 129
perspective 45, 58, 77, 125–31, 182n.5; false 188, 229
Peru 66
Phenomenology xix, 194–5, 201
Phidias 74
philosophy xv, 125
photograph, Alinari 76, Anderson 78
phronesis 4, 24
Picabia, Francis, *LHOOQ* 209
Picasso, Pablo 204
picturesque 87, 226, 228, 230
place x, 12, 22, 30–1, 38–9, 94, 132, 173, 175, 187, 196–9, 235; annihilated place 67; geometrising 126, 129–33; meeting 91, 207
Plato 5, 44, 120, 163, 166, 178; Academy 23; *The Dialogues* 178; *Phaedo* 155; Platonic idea 160, 163; *Republic* 160; *Sophist* 37, 41
Plotinus 164, 166–7; *Enneads* 160, 164–5
pluralism 49
Pluth, Ed 176, 179
poēsis 23, 99
poetics 179
Polizzotti, Mark 206
Pope, Alexander 46
positivism 87, 121, 133, 194, 200, 201, 204

261

Index

post-functionalist architecture 86–7
post-humanist 51, 81, 86–7, 89
post-industrial 50
postmodern xv, xx, 22, 29, 31, 50, 71, 95, 236
post-Second World War 64
praxis 3–4, 23–4, 25n.2, 136, 138–140, 143, 195–6
pre-modern 62
primitive 66, 201, 203–4, 209, 233
Prince Henry the Navigator 45
Prince Regent 46
psychoanalysis 10, 179, 183
Ptolemy 126–7, 129
Punch Drunk Theatre Company 99

Quarenghi, Giacomo 230
Queen Christina of Sweden 45

Ramelli, Agostino, *The Various and Ingenious Machines of Agostino Ramelli* 82
Ray, Man 203
recollective fantasia 173
Reductivism xv, 73, 131, 133, 201, 235
Renaissance xvii, xix, 45, 47, 67, 83–4, 89, 119–20, 125–9, 134, 137, 140, 147–159, 214–6, 242
Rentschler, Lothar 98
representation 120, 131, 139, 140, 143, 147–159, 187
Richter, Cornelia 239
Ricoeur, Paul 31
Rio de Janeiro 49
Risner, Friedrich 126
Robert, Hubert 233
Romania 45
Romano, Giulio, Palazzo del Tè 161
romanticism 4, 46, 193, 201, 204, 211, 230
Rome xiv, 45, 47, 68, 74, 76, 79, 105–7, 120, 160, 164, 166; Accademia di San Luca 120, 160–61 ; Arch of Constantine 76–8; Campidoglio 112; Colosseum 76; Pyramid of Cestius 76
Rossi, Aldo 218–9, 224; *The Analogical City* 222; *L'Architecture assassinée* 219, 222–3; *Architecture of the City* 221
Rossi, Carlo 230
Rousseau, Henri, *Still Life with Cherries* 211
Rousseau, Jean-Jacques 67, 229; *Julie, or the New Heloise: Letters of Two Lovers who Live in a Small Town at the Foot of the Alps* 228
Roussel, Raymond *How I Wrote Certain of my Books* 84
Impressions of Africa 84
Royal Society 62
Ruskin, John 75
Seven Lamps of Architecture 75
Russia 74, 226; Catherine the Great 230; Pavel Petrovich, Maria Feodorovna (wife of) 230; Pavlosk, Pavlovsk Park 226–236; Peter the Great 45; St Petersburg 45, 226, 230; Slavianka River 230; USSR 234

Said, Edward W. 71
Saint Pierre at Beauvais 62, 69n.25
Salis, Meta von 238–9
Sangallo, Antonio Da 157
Sassen, Saskia 5, 43, 48
Saussure, Ferdinand de 94
Sawday, Jonathan 83
Scandinavia 226
Scarpa, Carlo 212
Schelling, F. W. J. 241
Schiller, Friedrich 238–9, 241
Schleiermacher, Friedrich Daniel Ernst 193

Schopenhauer, Arthur 241
science xiv, 3, 125, 187, 189–90, 194; history of 129; modern xv; natural xv; physics 23, 127, 179, 191
Scottish Tay bridge disaster 62
Secession 239
Second Coming 45
Second World War 234, 245
Semper Gottfried 56, 75, 104, 194; *Der Stil* 194
sensus communis 30, 193
Serlio, Sebastiano, *Il secondo libro di prospettiva* 104
Shakespeare, William 7–8; *As You Like It* 67; *The Tempest* 67
Shelley, Mary 55; *The Last Man* 67
Shelley, Percy Bysshe 172
 Ozymandias 66
simulacra 13, 14
Sinclair, Iain 56, 92, 95, 99; *Dining on Stones* 95
Slingerland, E. G. xv
Smith, Horace 66
Snow, C. P., *Two Cultures* 172, 189
space, architectural 119, 128, 131, 136; geometrical 138; mental 141; situational 202, 212
space shuttle, *see also* Challenger 60
Spencer, Edmund 66
Spengler, Oswald 66; *Man and Technics* 65
Stalin, Joseph 234
Steiner, George 7, 50
Steiner, Rudolph 238
Stewart, George R., *Earth Abides* 66
Stonehenge 56, 105
structuralism 97
subjectivity 121, 173, 180
Surrealism, Surrealist(s) 199–212; symbolic order 101, 124, 178

tabula rasa 33
Tainter, Joseph 66

Tafuri, Manfredo 43, 214–225; *Architecture and Utopia (Progetto e utopia)* 215, 219; *L'Histoire assassinée* 188; *History of Modern Architecture, see also* Dal Co 43; *Ricerca del Rinascimento* 216; *The Sphere and the Labyrinth* 215; *Storia dell'architettura italiana 1944–1985* 222; *Theories and History of Architecture (Teorie e storia dell'architettura)* 215, 218, 222
Taylor, A. J. P. 64
technē 3, 24, 37, 41
technological, collapse 67; decline 55; failure 62; pessimism 61
technology xv, 3, 55, 132; age of xiv; spectre of 55
Terry, Quinlan 87
Texas 107
Thames 99–100; Estuary 56; Gateway 56, 91–2, 94, 97, 101
Thamesmead 96
Theodoric of Fribourg, *De iride et radialibus impressionibus* 126
theology xv
Theory in Landscape Architecture 226
theoria 23
Thompson, Jon 88
Thoreau, Henry David 107
Todtnauberg 21
topos 130
tourism 47
transitivity 180, 184n.18
Tzonis, Alexander 12

United States of America 61, 66, 217; Albemarle County 111; Cape Cod 107; Carr's Hill 112; Cincinatti 107; Houston 107, 111; New York City 115
University of Basel 237
University of Greenwich 92
University of Innsbruck 92

Index

University of Pennsylvania Press 226
urban form 46, 222
urban planning 50

Vaccaro, Domenico Antonio 175
Vænius, Otto 174
Van de Velde, Henry 186, 237–246
Vatican 147, 149; Saint Peter's Basilica 78, 120, 147–59
Vaughan, Diane 60
Venice 100; Biennale 81; Institute of Architecture of the University of Venice (IUAV) 214, 218–9
Verene, Donald Phillip 178
Veronica's veil 156
Vesely, Dalibor 29, 137
Vico, Giambattista 30, 121, 172–184, 193; *Autobiography* 178–9, 183; Cebes' Table 174, 179–80; *The New Science* 121, 173–5, 178–80, 183
Victorian 62
Vienna 45, 203; Vienna University of Technology 92
Virgil 7, 55, 68; Aeneid, The 104
Virginia 107, 111
Virilio, Paul 55; *The Original Accident* 58
Vita activa xvii
Vita contemplativa xvii, 241

Vitruvius xiv, 44, 83
Voegelin, Eric; *Order and History* 4, 44
Volvo 95

Wagner, Richard 238, 241
Walden, Henry David Thoreau 105
Wacjman, Judy 59–60
Watt, James 59
White, Hayden 31
Whitehead, Alfred North 49
Whitehead, Neil H. 5
Witelo, *Perspectiva* 126
Wittgenstein, Ludwig 56, 92, 96, 98; *Tractatus Logico-Philosophicus* 98; sisters Hermine and Margarethe 98
World Health Organization 57

The X Files 172

Yates, Frances 83; *The Art of Memory* 97

Zevi, Bruno; *Storia dell'architettura moderna* 218
Zuccari, Federico 167–9; *L'Idea de' pittori, scultori ed architetti* 120, 160–61, 162
Zurich 114

eBooks – at www.eBookstore.tandf.co.uk

A library at your fingertips!

eBooks are electronic versions of printed books. You can store them on your PC/laptop or browse them online.

They have advantages for anyone needing rapid access to a wide variety of published, copyright information.

eBooks can help your research by enabling you to bookmark chapters, annotate text and use instant searches to find specific words or phrases. Several eBook files would fit on even a small laptop or PDA.

NEW: Save money by eSubscribing: cheap, online access to any eBook for as long as you need it.

Annual subscription packages

We now offer special low-cost bulk subscriptions to packages of eBooks in certain subject areas. These are available to libraries or to individuals.

For more information please contact webmaster.ebooks@tandf.co.uk

We're continually developing the eBook concept, so keep up to date by visiting the website.

www.eBookstore.tandf.co.uk